Luay AlSwidi
Thu-al-fiqar Al-aamery

Resolvability In Soft Topological Spaces

Luay AlSwidi
Thu-al-fiqar Al-aamery

Resolvability In Soft Topological Spaces

Soft Topological Spaces

Noor Publishing

Imprint
Any brand names and product names mentioned in this book are subject to trademark, brand or patent protection and are trademarks or registered trademarks of their respective holders. The use of brand names, product names, common names, trade names, product descriptions etc. even without a particular marking in this work is in no way to be construed to mean that such names may be regarded as unrestricted in respect of trademark and brand protection legislation and could thus be used by anyone.

Cover image: www.ingimage.com

Publisher:
Noor Publishing
is a trademark of
Dodo Books Indian Ocean Ltd., member of the OmniScriptum S.R.L Publishing group
str. A.Russo 15, of. 61, Chisinau-2068, Republic of Moldova Europe
Printed at: see last page
ISBN: 978-620-4-72046-3

Resolvability In Soft Topological Spaces

By

Dr. Luay Abd Al-Haine Al-Swidi

Thu-al-fiqar Faker Naseef Al-aamery

Contents

Introduction 5

Chapter One: Soft set theory

Introduction..13

1.1 Soft sets 13

1.2 Soft functions 37

Chapter Two: soft topology

Introduction.. 44

2.1 Soft topology..43

2.2 Soft (limits , boundary , exterior) points....................58

Chapter Three: soft resolvable space

Introduction..71

3.1 Density in soft topological spaces................................71

3.2 Soft resolvability and soft irresolvability 83

Refrences.. 94

List of symbols

Symbols	Description
χ	The universal set
E_χ	The set of all possible parameters
$IP(\chi)$	The power set of χ
F_A	Soft set
$S_i(\chi)$	The family (i) i=1,2,3,4
$\neg A$	The negation of a set A
$\widetilde{\chi}_A$	The absolute soft set
$\widetilde{\Phi}_A$	The null soft set
$\widetilde{\chi}_A - F_A$	The soft complement of F_A
e_F	The soft point of F_A
x_e	The a $-$ sigleton soft point
x_a	The a $-$soft point

Symbols	Description
$\widetilde{\mathcal{X}}_a$	The absolute soft set with respect to a
$\widetilde{\Phi}_a$	The null soft set with respect to a
$\widetilde{\cap}$	The soft intersection
$\widetilde{\cup}$	The soft union
Symbols	**Description**
$\widetilde{\cap}_a$	The $a-$soft intersection
$\widetilde{\cup}_a$	The $a-$soft union
$C\ell$	The soft Closure
Sdh	The soft cluster
$C(\widetilde{\mathcal{X}})$	The set of all soft closed set
Sb	The soft boundary
D	The soft derive
ext	The soft exterior
$a-Cl$	The $a-$soft closure
$a-Int$	The $a-$soft interior
$a-Sb$	The $a-$soft boundary
$a-D$	The $a-$soft derive
f_{pu}	The soft map
Iff	if and only if
Int	The soft interior
$|F_A|$	The cardinality of F_A

Introduction

Molodtsov [1] started with the soft set theory as a mathematical tool to deal with cases that have some kind of a doubt (uncertainty) , usually the normal mathematical tools are unable to find a solution of them . By putting in consideration the real need to find one word solutions to these problems . This scientist introduces a lot of tools to solve a great number of social problems that have a great relation with nans life ; directly or indirectly .

Molodtsov [1] defines the soft set , and the majority researchers follow his paces , one of them , P.K. Maji in 2003 [2] and M.Ifran Ali in 2009 [11] , the definition of these researchers and who follows them of a soft set depends on three tools :

   i. E , the set off all possible parameters .

  ii. χ , the universal set that includes things that has a relationship with the parameters .

 iii.   The map F , which is from E to $IP(\chi)$ [the power set of χ] from defined as the following :

       If $A \subseteq E$, a pair (F,A) is called a **soft set** , where F is a mapping defined as $F : A \longrightarrow IP(\chi)$, consequently , the family of all soft sets that were defined on $IP(\chi)$ are as the following :

$$S_1(\chi) = \left\{ F_{i_{A_i}} , A_i \subseteq E , F_i : A_i \longrightarrow IP(\chi) \right\}.$$

In (2011) Naim cagman[4] , S.Karatas and S.Eningolu defines another type of a soft sets , as follows :

$$S_2(\chi) = \left\{ F_{A_i} , A_i \subseteq E , F : A_i \longrightarrow IP(\chi) \right\} \text{ , that is they change the sets of}$$

parameters .

In (2012) S.Atmagn , M.Akdag , I.Zorolutuna and N.K.Min [31] defines a new type of a soft sets . It depends on the following formula :

$$S_3(\chi) = \{F_{i_A} , A \subseteq E , F: A \longrightarrow Ip(\chi)\} .$$

Researches have been developed three types of a soft sets . A certain "mood" is achieved by some of these types not by all , this is what will be discussed in this work .

In general , the main concepts are the same in all these types , excepts soft union , soft intersection and soft complement . they will be clarified in what follows .

G. Xuenchong , X [32]. presents a new type soft sets known as "Central soft sets" . It depends on a fundamental set $(F: E \longrightarrow IP(X))$, the central soft set stands for a restricted of the map F to the subset A of E . This type differs from all other types in that it deals with the map and its domain and its codomain in the same time . For example in [30] , it is defined the " soft complement of a soft set " as follows :

$(F, A)^C = (F^C, A^C)$, $F: A \longrightarrow IP(X)$. The difference , of course , is fundamental .Accordingly , from the topological point of view , there will be four types of soft topological spaces each of them depends on the family that contains the soft topological space .

In [2] example (2.6) doesn't stand for an accurate one .It is believed that the soft set G_B doesn't necessarily be a soft set due to Molodtsov's definition .This is so because the set $(\neg A)$ forms a subset of the set of parameters E_X .When this is the case , it can form a domain of the relation (F) and form (F, A) due to a soft set from the set E_X. This set E_X ' in turn' is unlimited since points out that it represents characteristics , properties and symbols .Characteristics and properties are unidentifiable facts . The negation of the elements of this subset from E_X doesn't mean that the rest is negation of these elements .

In other words , we can't determine the unlimited facts that represent $(\neg A)$ and hence this expression is inaccurate .

As an example , if the universal set χ contains a set of pictures as $(A \subseteq E_\chi)$ and it contains several parameters as (sad , angry and proud) , the negation of the adjective (sadness) will be (happiness) and the negation of the adjective of (anger) will be (quietness) and the same with the adjective of (pride) it will be (humbleness) .

From the above example , we want to say that each parameter should have a direct antonym and it is accurately the opposite . This note has lost the focus of several researches in spite of the fact that it represents basic factors in drawing a soft set . There are problems appeared as we have indicated in example (17) from [P . k.Maji , R . Biswas and A . R . Roy (2003)] , the resource [S.M.Modumugal pal (2013)] and example (6) , the research falls in the same problem and in the resource [P.k.Maji , and A.R.Roy (2002)] as well .

On this statue , we opine to make E_χ by the following expression :

$E_\chi = E \cup \neg E$, where $\neg E$ is the negation set of E , this means that E and , $\neg E$ are disjoint which implies that :

$E_\chi \backslash E = \neg E$ and $E_\chi \backslash \neg E = E$.

By taking in consideration the last period , the care in this field has a gained many development , and the voice of this field spread in rapid range, because of the light nature that it stand on soft sets .In [P.k.Maji etal (2001), N . Cagman ,F . Cltak and S . Enginolu (2001)] maji et al had studied [soft set theory] and used this theory in

several problems that had a relation in making decision . In addition to that he presented a concept [fuzzy soft set] and generalized several concepts in [fuzzy soft set] , and he studied it properties as well .In [M . I .Ali etal (2009)] [11], Ali etal defined some of new operations in soft set , in [P . k .Maji , R . Biswas and A .R. Roy (2003)] [2] Aktas and Cageman wrote in soft sets and soft groups [4] , in (2008) , Z.Kong et al and commanton (2009)] Kong etal also applied the soft set theory in problem relations making decision .

The previous studies :-

- In 1999 , Molodtsov [1] begins whit the soft set theory as mathematical tool to solve many complicated problems in economics , engineering , and environment which we cannot successfully use classical methods because of various uncertainties typical of those problems .There are three theories: theory of probability, theory of fuzzy sets, and the interval mathematics which we can consider as mathematical tools for dealing with uncertainties. But all these theories have their own difficulties.
  Theory of probabilities can only deal with stochastically stable phenomena , Without going into mathematical details .
  Interval mathematics have arisen as a method of taking into account the errors of calculations by constructing an interval estimate for the exact solution of a problem. This is useful in many cases, but the methods of interval mathematics are not sufficiently adaptable for problems with different uncertainties. They cannot appropriately describe a smooth changing of information, unreliable, not adequate, and defective information, partially contradicting aims, and so on. Into mathematical details .

- In 2003 P. K.Maji [2] , R. Biswas and A .R.Roy studied a soft set theory and they introduced many of new concepts of this theory as an inclusion relations between the soft set , the formula of the empty set in this theory , the equality of two soft sets , the complement of a soft set also the soft intersection and the soft union with some of important results and properties .

- N.Cagman , Serkan Karatas ,Serdar Enginogllu [4] studied the soft topology after the success of Molodtsov in appling the soft set theory in several directions , such as the smoothness of functions , game theory , operations research , Riemann integration , perron integration , probability , theory of measurement and so on . They are introduced a new type of a soft set and a new definition of a soft intersection and a soft union with many results and properties .

- M.ifran Ali , F.feng ,X. liu W.Keun min and M. shabir [11] discussed new operations in the soft set theory such as de_ morgans laws in the soft set theory in addition to some anew views of the soft union and the soft intersection .
- P.Majumdar and S.K . Samanta introduced a generalized of a fuzzy soft sets [12].
- H.Aktas and N.Cageman studied the soft sets and soft groups [20] .
- Z.Kong et al , comment on " A fuzzy soft set theoretic approach to decision making problems [22] .
- N,Cageman and S.Enginolgu intrudused a new study of a soft set theory and uni-int decision making European journal of operational research .

Research content :-

This thesis consists of three chapters :

In chapter one , we introduced two sections . In section one we recorded some of the concepts in the soft set theory (definitions and properties) as well as previous theorems that we need in our work . and , a comparison between the families of a soft sets which is intruduced by a previous studies .

In section two we are introduced a new concept of a soft function with some of remarks and important properties .

Chapter two consists of two sections . Section one consists of a soft topology with some of new definitions in addition to a subspace in soft set theory and the soft continuity .

In section two , we introduced new definitions of many soft topological concepts , such as soft limits points , soft isolated points , soft boundary points and soft exterior points whit there properties and the relationship theorems .

Chapter three contains also two sections . In section one we put a new concept of the dense sets in soft topological spaces depending on the soft closure with some of theorems which are needed in section two . In section two we are studied the resolvability and irresolvability in soft topological spaces .

***Objective of study :-**

We have many ways to solve our problems as follows :

1. To an analytical study on the soft set theory and classify the types of soft sets as four families and explaining the deference between its properties .
2. To define a new type of the concept of density in the soft set theory and studying its properties .
3. To introduce the concept of the resolvability and irresolvability in the soft set theory and its other concepts related to them such as , soft proportional resolvable , soft proportional irresolvable , soft strongly irresolvable , soft submaximal and other concepts .

Chapter one

Soft Set Theory

1-Introduction

In this chapter , we will study some basic notions of the soft set theory , such as : soft set , soft intersection , soft union , soft complement , soft deference ...Etc . We had a comparison between the types of soft sets and study each of its properties . So we need to define some of the important characteristics :

(1.1) Soft sets

In this section , we divided the soft sets into four families in order to explain the difference between each of them , in addition to some notes .The following concepts represent the basic tools of a soft set .

i) χ be the initial universe set and E_χ be the set of all possible parameters under consideration with respect to χ.

ii) $IP(\chi)$ is denoted to the power set of χ, (i . e the family of all subsets of χ) .

iii) $A \subseteq E_\chi$.

Usually parameters are attributes ,   characteristics or properties of objects in χ, also for E_χ may simply be denoted by E [4] always means the universal set of parameters with respect to χ, unless otherwise specified .

Definition1.1 .1 [2]

A pair (F, A) is called a soft set over χ , where F is a mapping defined as : $F: A \longrightarrow IP(\chi)$, in other words , a soft set over χ is a parametrized family of a subset of the universe χ , for $a \in A$, $F(a)$ may be considered as the set of a- approximate elements of the soft set (F, A) . We will denote simply by F_A of the pair (F , A).

❖ The soft set $F_A = (F, A)$ is of the form $F_A = \{(a, F(a)), a \in A\}$, also we denoted the elements of the soft set F_A by $\alpha = F_a = \{(a, F(a))\}$ for $a \in A$.

Example1.1.2

Suppose that there are three houses in the universe χ, $\chi = \{h_1, h_2, h_3\}$ under consideration and $A \subseteq E_\chi$, such that $A = \{x_1, x_2\}$, is a subset of the decision parameter x_i, $i = 1,2$, stand for the parameters (wooden , expansive) , respectively , consider the mapping F_A given by "houses(.) , " where (.) is to be filled in by one of the $x_i \in A$.

For instance , $F_A(e_1)$ means "houses (wooden) " and its functional values is the set $\{h \in \chi: h \text{ is a wooden huose}\}$.

Let $F(x_1) = \{h_1, h_2\}$, $F(x_2) = \{h_3\}$, then we can view the soft set F_A as consisting of the following collection of approximations :

$$F_A = \{(x_1, \{h_1, h_2\}), (x_2, \{h_3\})\}.$$

Definition 1.1.3

Let A be a non-empty set , the NOT set of A is denoted by $\neg A$ is defined by : $\neg e = not\ e \quad \forall\ e \in A$

Example 1.1.4

Let $A = \{\text{expansive , cheap , beautiful}\}$

$\neg A = B = \{\text{not expansive , not cheap , not beautiful}\}$

Proposition 1.1.5 [2]

For any subsets A, B of E_χ the following properties are holds:

i) *If $A \subseteq B$ then $\neg A \subseteq \neg B$*

ii) *If $a \in A$ then $\neg a \in \neg A$*

iii) $\quad \neg(\neg A) = A$

iv) $\neg(A \cup B) = \neg A \cup \neg B$

v) $\neg(A \cap B) = \neg A \cap \neg B$

vi) $\neg(A \times B) = \neg A \times \neg B$

Corollary 1.1.6

For any non-empty sets $A, B \subseteq E_\chi$ and index set $\wedge$ we have :

i) $\neg\left(U_{\lambda \in \wedge} A_\lambda\right) = U_{\lambda \in \wedge} \neg A_\lambda$

ii) $\neg\left(\cap_{\lambda \in \wedge} A_\lambda\right) = \cap_{\lambda \in \wedge} \neg A_\lambda$

iii) $\quad \neg(A - B) = \neg A - \neg B$

Proof : (i)

Since $A_\lambda \subseteq U_{\lambda \in \wedge} A_\lambda$ so by proposition -5- part (i) we have :

$U_{\lambda \in \wedge} \neg A_\lambda \subseteq \neg\left(U_{\lambda \in \wedge} A_\lambda\right)$ ....(1.1)

Now let $x \in \neg\left(U_{\lambda \in \wedge} A_\lambda\right)$, so $\exists\, a \in U_{\lambda \in \wedge} A_\lambda$ such that : $x = \neg a$, since

$a \in U_{\lambda \in \wedge} A_\lambda$, $\exists\, \lambda_0 \in \wedge$ such that $a \in A_{\lambda_0}$ and $x = \neg a \in \neg A_{\lambda_0} \subseteq$

$U_{\lambda \in \wedge} \neg A_\lambda$, hence from (1.1), $\quad \neg\left(U_{\lambda \in \wedge} A_\lambda\right) = U_{\lambda \in \wedge} \neg A_\lambda$.

Proof : (ii)

Since $\cap_{\lambda \in \wedge} A_\lambda \subseteq A_\lambda$, so by proposition -5- part (i) we have:

$\neg\left(\cap_{\lambda \in \wedge} A_\lambda\right) \subseteq \cap_{\lambda \in \wedge} \neg A_\lambda$ .... (1.2)

now $\quad$ Let $x \in \cap_{\lambda \in \wedge} \neg A_\lambda$, thus $x \in \neg A_\lambda \quad \forall\, \lambda \in \wedge$ imply that

$\neg x \in \neg(\neg A_\lambda) = (A_\lambda) \quad \forall\, \lambda \in \wedge \quad$ [by proposition -5- part -3-]

so $\neg x \in \cap_{\lambda \in \wedge} A_\lambda$, thus $x = \neg(\neg x) \in \neg\left(\cap_{\lambda \in \wedge} A_\lambda\right)$ .... (1.3))

from (1.2) and (1.3) we have :

$$\neg\left(\cap_{\lambda \in \wedge} A_\lambda\right) = \cap_{\lambda \in \wedge} \neg A_\lambda$$

Proof : (iii)

$x \in \neg(A - B)$ if and only if $x = \neg a$ for some $a \in A$ and $a \notin B$ if and only if $\neg a \in \neg A$ and $\neg a \notin \neg B$ if and only if

$x \in \neg A - \neg B$ hence :

$$\neg(A - B) = \neg A - \neg B .$$

Example 1.1.7 [2]

Suppose that there are five wooden houses in the universe χ and E_χ be a set of all possible parameters given by :

$$\chi = \{h_1, h_2, h_3, h_4, h_5\}$$

Let $A \subseteq E_\chi$ such that $A = \{$ brick , muddy , steel , stone $\}$ be the set of parameters showing the building of the houses ,

Let B be the NOT set of the set A , that :

$B = \neg A = \{$ not brick , not muddy , not steel , not stone $\}$

Let F_A and G_B are two soft sets over the same universe χ , which describe the construction of the houses , the soft sets F_A and G_B can be regarded as a collection of approximation , that is :

$F_A = \{$ brick houses $= \emptyset$, muddy houses $= \emptyset$, steel houses $= \emptyset$, stone houses $= \emptyset\}$,

It is easy since all of them are wooden , and

$G_B = \{$not brick houses $= \chi$, not muddy houses $= \chi$, not steal houses $= \chi$, not stone houses $= \chi\}$.

Also it is easy since all them are wooden .

Discussion :

From the above example we want to say that we believe the set G_B is not (soft set) according to [definition 1.1.1] , because the fact that must be hold , [the set $\neg A$ is a subset of E_χ] , if it is so , then it can be a domain of the function F , and then F is a soft set , but actually , the set $(\neg A)$ is a (mysterious) set and non-specific parameters , [since E_χ is

Non-specific parameters] , as we a indicated that it is usually (characteristics or properties) , and these facts cannot be determined , and a subset of E doesn't mean that the other parameter is the negative of the subset . [by the other hand , we cannot define which of the parameters that are undefined number indicates to $(\neg A)$, and consequently , this expression will not be accurate at all . As an example , if the universal set χ contains a set of pictures as $(A \subseteq E_\chi)$ it contains several parameters as (sad , angry , proud) , the negation of the adjective (sadness) will be (happiness) and the negation of the adjective of (anger) will be (quietness) and the same with the adjective of (pride) it will be (humbleness) .

From the above example , we want to say that each parameter should have a direct antonym and it is accurately the opposite . This note has lost the focus of several researches in spite of the fact that it represents basic factors in drawing a soft set . there are problems appeared as we have indicated in example (17) from [10] , the resource [15] and example (6) , the research falls in the same problem and in the resource [13] as well .

On this statue , we opine to make E_χ by the following expression :

$E_\chi = E \cup \neg E$, where $\neg$ E is the negation set of E , this means that E and , $\neg$ E are disjoint which implies that :

$E_\chi \backslash E = \neg E$ and $E_\chi \backslash \neg E = E$. we called $\neg E$ is the negation of E_χ , and this mean that E and $\neg$ E are disjoint .

Families of the soft sets :

In this section of the research , we will introduce all the types of soft sets that we are called (families of a soft sets) with explaining the properties of each type , each family is different than the other in some adjectives and properties and agrees with the other , as in the following :

First family

We will denote this family as $S_1(\chi)$ and it contains the following type :

$$S_1(\chi) = \{ F_{i_{A_i}}, \text{ where } A_i \subseteq E_\chi \text{ and } F_i: A_i \longrightarrow IP(\chi) \}.$$

For example , if χ is any universal set and E_χ is the set of all parameters , if $A , B, C \subseteq E_\chi$, then $F_A , G_B, H_C \in S_1(\chi)$.

In order to know the properties of this family we must write some of important definitions to illustrate these properties , as the following :

Definition 1.1.8 [1]

The soft union of any two soft sets F_A and G_B is a soft set H_C , where $C = A \cup B$ and $\forall c \in C$:

$$H_C(c) = \begin{cases} F_A(c) & if c \in A \setminus B \\ G_B(c) & if \ c \in B \setminus A \\ F_A(c) \cup G_B(c) & if \ c \in A \cap B \end{cases}$$

This is denoted by $F_A \, \widetilde{U} \, G_B = H_C$ or by ,

$$F_A \, \widetilde{U}_E \, G_B = H_C .$$

Definition 1.1.9 [2]

The soft intersection of two soft sets F_A and G_B is the soft set H_C where $C = A \cap B$ and $\forall c \in C$ and define :

$$H_C(c) = F_A(c) \cap G_B(c)$$

We write $F_A \, \widetilde{\cap} \, G_B = H_C$

Definition 1.1.10[2]

For a soft set F_A over the universe χ, the relative complement of F_A is denoted by $(F_A)^{\,c} = F_A^{\,c}$ where $F^c: A \longrightarrow IP(\chi)$ is a mapping given by $F^c(\alpha) = \chi - F(\alpha) \; \forall \, \alpha \in A$.

Definition 1.1.11 [25]

The restricted union of F_A and G_B is defined as the soft set H_C such that :

$$F_A \, \widetilde{U}_R \, G_B = H_C \quad , \text{where} \quad C = A \cap B \text{ and } \forall c \in C :$$

$$H_C(c) = F_A(c) \cup G_B(c).$$

Definition 1.1.12 [36]

Let χ be the initial universe and let E_χ be the set of parameters , let F_A and G_B be two soft sets over the common universe χ , and $A, B \subseteq E_\chi$, then F_A is a soft subset of G_B denoted by $F_A \widetilde{\subseteq} G_B$ if :

i) $A \subseteq B$

ii) for all $a \in A$, $F(a)$ and $G(a)$ are the identical approximation .

Definition 1.1.13 [2]

A soft set F_A over the universe χ is called a null soft set and denoted by $\widetilde{\Phi}$ if $\forall\, a \in A$, $F(a) = \varphi$

That is : $\widetilde{\Phi} = \{(a, \varphi)\}; a \in A\}$.

Definition 1.1.14 [25]

A soft set F_A over the universe χ is called an absolute soft set and denoted by $\tilde{\chi}$, if $\forall a \in A$, $F(a) = \chi$

That is : $\tilde{\chi} = \{(a, \chi)\}; a \in A\}$.

Example 1.1.15

If the universe set $\chi = \{c_1, c_2, c_3\}$ and $A = \{e_1, e_2, e_3\}, B = \{e_2, e_3\}, C = \{e_1, e_3\}$,let soft set $F_A = \{(e_1, \chi), (e_2, \chi), (e_3, \chi)\}$, $G_B = \{(e_2, \{c_1, c_2\})), (e_3, \{c_1\})\}$ and $H_C = \{(e_1, \{\varphi\})), (e_3, \{\varphi\})\}$ then :

$$F_A \,\widetilde{\cup}_E\, G_B = \{(e_1, \chi) , (e_2, \chi), (e_3, \chi)\}$$
$$F_A \,\widetilde{\cap}\, G_B = \{(e_2, \{c_1, c_2\})), (e_3, \{c_1\})\} .$$

$$F_A = \{(e_1, \chi), (e_2, \chi), (e_3, \chi)\} = \tilde{\chi}$$

$$F_A{}^C = \chi - F(e) \quad \forall\, e \in A$$

$$= \{(e_1, \varphi), (e_2, \varphi), (e_3, \varphi)\} = \widetilde{\Phi} .$$

proposition1.1.16 [4]

let F_A any soft set over the universe χ , then the following are hold :

i) $F_A \,\tilde{\cup}\, F_A = F_A$
ii) $F_A \,\tilde{\cap}\, F_A = F_A$
iii) $F_A \,\tilde{\cup}\, \tilde{\Phi} = F_A$
iv) $F_A \,\tilde{\cap}\, \tilde{\Phi} = \tilde{\Phi}$
v) $F_A \,\tilde{\cup}\, \tilde{\chi} = \tilde{\chi}$
vi) $F_A \,\tilde{\cap}\, \tilde{\chi} = F_A$

Second family:

We will denoted $S_2(\chi)$ and it contains the following type :

$$S_2(\chi)=\{ F_{A_i} : A_i \subseteq E_\chi \text{ and } F: A_i \longrightarrow IP(\chi)\}$$

We must write some of the properties and definitions of this family .

Definition 1.1.17 [4]

A soft set F_A on the universe χ is denoted by the set of ordered pairs ,

$$F_A =\{ (x, f_A(x)), x \in E_\chi, f_A(x) \in IP(\chi) \}$$

Where $f_{A:} E \longrightarrow IP(\chi)$, such that $f_A(x) = \phi \;\forall\; x \notin A$

Here f_A is called an approximate function of the soft set F_A , the value of $f_A(x)$ may be arbitrary , some of them may be empty , some may have non-empty intersection .

Definition 1.1.18 [4]

let F_A , F_B are two soft sets over χ then soft intersection and the soft union between them are defined by the approximate functions :

$$f_{A\cup B}(x)=f_A(x) \cup f_B(x), f_{A\cap B}(x) = f_A(x) \cap f_B(x).$$

Respectively the soft complement of F_A is defined by:

$f_{A^{\tilde{c}}}(x) = f^c{}_A(x)$, where $f^c{}_A(x)$ is the complement of the set $f_A(x)$, that is:

$f^c{}_A(x) = \chi - f_A(x)$ for all $x \in E_\chi$.

Definition 1.1.19 [4]

Let $F_A \in S_2(\chi)$, if $f_A(x) = \varphi$ for all $x \in E$, then F_A is called an empty set, and denoted by $\tilde{\Phi}$.

Definition 1.1.20 [4]

Let $F_A \in S_2(\chi)$, if $f_A(x) = \chi$ for all $x \in A$, then F_A is called A-universal soft set, and denoted by $\tilde{\chi}$. If $A=E$ then E-universal soft set is called universal soft set denoted by $\tilde{\chi}$.

Definition 1.1.21 [4]

Let $F_A, F_B \in S_2(\chi)$, then F_A is called a soft sub set of F_B and denoted by

$F_A \tilde{\subseteq} F_B$ if $f_A(x) \subseteq f_B(x)$ for all $x \in E$.

proposition 1.1.22 [4]

Let $F_A \in S_2(\chi)$ then :the following statements are hold

i) $F_A \tilde{\cup} F_A = F_A$, $F_A \tilde{\cap} F_A = F_A$

ii) $F_A \tilde{\cup} \tilde{\Phi} = F_A$, $F_A \tilde{\cap} \tilde{\Phi} = \tilde{\Phi}$

iii) $F_A \tilde{\cup} \tilde{\chi} = \tilde{\chi}$, $F_A \tilde{\cap} \tilde{\chi} = F_A$

iv), $F_A \tilde{\cup} F^c{}_A = \tilde{\chi}$, $F_A \tilde{\cap} F^c{}_A = \tilde{\Phi}$

Example 1.1.23

See example 1.1.15

proposition 1.1.24 [4]

Let $F_A, F_B, and\ F_C \in S_2(\chi)$ then,

i) $F_A \tilde{\cup} F_B = F_B \tilde{\cup} F_A$, $F_A \tilde{\cap} F_B = F_B \tilde{\cap} F_A$

ii) $(F_A \tilde{\cup} F_B)^C = F^c{}_A \tilde{\cap} F^c{}_B$, $(F_A \tilde{\cap} F_B)^C = F^c{}_A \tilde{\cup} F^c{}_B$

iii) $(F_A \tilde{\cup} F_B) \tilde{\cup} F_C = F_A \tilde{\cup} (F_B \tilde{\cup} F_C)$

iv) $(F_A \tilde{\cap} F_B) \tilde{\cap} F_C = F_A \tilde{\cap} (F_B \tilde{\cap} F_C)$

v) $(F_A \tilde{\cap} F_B) \tilde{\cup} F_C = (F_A \tilde{\cup} F_C) \tilde{\cap} (F_B \tilde{\cup} F_C)$

vi) $F_A \tilde{\cap} (F_B \tilde{\cup} F_C) = (F_A \tilde{\cap} F_B) \tilde{\cup} (F_A \tilde{\cap} F_C)$

Third family

Central soft sets :

We will denote $S_3(\chi)$, and it is defined as follows :

$$S_3(\chi) = \{f_{i_{A_i}} , \text{where } A_i \subseteq E_\chi \text{ and } f_i: E_\chi \longrightarrow Ip(\chi)\} .$$

Definition 1.1.25 [32]

Let A be a subset of the parameter set E , a pair (f, A) is called a soft set with a central set A , where f is a mapping given by $f: E \longrightarrow IP(\chi)$,

For simply we call (f, A) as a central soft set .

For two central soft sets $(f, A) \; and \; (f, B)$, we say that

$$(f, A) = (f, B)$$

, if $A = B$ and $f = g$

" In fact a central soft set (f, A) results a complete parametrization of the universe set χ by the mapping f , the concept here is different from the previous [1,4] . The central parameter set is a central soft set , is to illustrate particularities which exist in information given by a soft set . These particularities can have a variety of meaning , with different back rounds of problems , for example in the instance of choosing houses , let (f, A) be a central soft set that present information about the scoring given by Mr. X , we can assume that A is A parameters set related with his field of expertise , to take another instance , teachers in school , who are good at teaching certain subjects and X consists of all teachers in this school , for each participating student X_i .

We assume that A_i is the set of his excellent courses , a mapping

$f_i : E \longrightarrow IP(\chi)$, denotes the evaluation results by X_i. [32]

Example 1.1.26.

Suppose that U = $\{h_1, h_2, h_3, h_4, h_5\}$ represents a set of houses. Let $E = \{I, II, III, IV, V, VI, VII, VIII\}$, where roman numerals I to VIII represent some
attributes of houses respectively such as reasonable design space, green environment,
excellent property management, convenient transportation. Consider two central soft
sets (f, A) and (g, B) defined over χ, where $A = \{IV, V\}$ and $B = \{I, IV\}$ are two
central sets which are precisely the areas of expertise of Mr. X and Mr. Y in choosing
houses respectively, the mappings f and g are defined as follows:
$f(I) = \{h_2, h_3, h_4, h_5\}, f(II) = \{h_2, h_5\},$
$f(III) = \{h_2, h_3\}, f(IV) = \{h_1, h_4, h_5\},$
$f(V) = \{h_1, h_4\}, f(VI) = \{h_1, h_5\},$
$f(VII) = \{h_2, h_5\}, f(VIII) = \{h_3, h_4, h_5\};$
and
$g(I) = \{h_2, h_3, h_4\}, g(II) = \{h_1, h_5\},$
$g(III) = \{h_3, h_4\}, g(IV) = \{h_2, h_4, h_5\},$
$g(V) = \{h_2, h_4\}, g(VI) = \{h_4, h_5\},$
$g(VII) = \{h_2, h_4\}, g(VIII) = \{h_1, h_4, h_5\}.$
By definitions of union and intersection on central soft sets, we have
$(f, A) \sqcup (g, B) = (h, \{I, IV, V\}),$ where h is defined as:
$h(I) = \{h_2, h_3, h_4\}, h(II) = \{h_1, h_2, h_5\},$
$h(III) = \{h_2, h_3, h_4\}, h(IV) = \{h_1, h_2, h_4, h_5\},$
$h(V) = \{h_1, h_4\}, h(VI) = \{h_1, h_4, h_5\},$
$h(VII) = \{h_2, h_4, h_5\}, h(VIII) = \{h_1, h_3, h_4, h_5\}.$
If we take a new central soft sets (f, C), where $C = \{III, IV\}$. Then
$(f, C) \sqcup (g, B) = (j, \{I, III, IV\}),$ where j is defined as:
$j(I) = \{h_2, h_3, h_4\}, j(II) = \{h_1, h_2, h_5\},$
$j(III) = \{h_2, h_3\}, j(IV) = \{h_1, h_2, h_4, h_5\},$
$j(V) = \{h_1, h_2, h_4\}, j(VI) = \{h_1, h_4, h_5\},$
$j(VII) = \{h_2, h_4, h_5\}, j(VIII) = \{h_1, h_3, h_4, h_5\}.$
Since we choose two different central sets for the mapping f, two different central
soft sets $(h, \{I, IV, V\})$ *and* $(j, \{I, III, IV\})$ are obtained. It demonstrates that central
sets play an important role as mappings in the operation of union.

The function of central soft sets is namely reflected in the superiority of operations between soft sets , which will be shown in the next definition. "

Definition 1.1.27 [32]

The union of two central soft sets $(f, A), (g, B)$ over a common universe χ is a central soft set (h, C) , where $C = A \cup B$, and for all $e \in E$

$$h_C(e) = \begin{cases} f_A(e) & \text{if } e \in A \setminus B \\ g_B(e) & \text{if } e \in B \setminus A \\ f_A(e) \cup g_B(e) & \text{other wise} \end{cases}$$

We write $(f, A) \sqcup (g, B) = (h, C)$.

This operation between soft sets   is   given from   the view of information   sciences ,   relationship between central soft sets will also be better , represented by these operations defined here .

Definition 1.1.28[32]

The intersection between two central soft sets (f, A) *and* (g, B) over a common universe χ is a central soft set (h, c) where $C = A \cap B$, and for all $e \in E_\chi$.

$$h_C(e) = \begin{cases} g_B(e) & \text{if } e \in A \setminus B \\ f_A(e) & \text{if } e \in B \setminus A \\ f_A(e) \cap g_B(e) & \text{other wise} \end{cases}$$

We write $(f, A) \sqcap (g, B) = (h, A \cap B)$.

This definition may seems strange , however , actually it gives the maximum central soft set which is contend in two original central soft sets .

Definition 1.1.29[32]

For two central soft sets (f,A) *and* (g,B) over a common universe χ , we write $(f,A) \sqsubseteq (f,B)$ and   call it a central soft information order if $(f,A) \sqcup (g,B) = (g,B)$.

proposition 1.1.30 [32]

For two central soft sets (f,A) *and* (g,B) over a   common universe χ , $(f,A) \sqsubseteq (g,B)$ iff the following conditions are true .

i) $A \subseteq B$

ii) $(f,A) \sqcap (g,B) = (f,A)$

Definition 1.1.31 [32]

The complement of   a   central soft   set (f,A) is denoted by $(f,A)^c = \left(f^c, A^c\right)$, where $A^c = E - A$, $f^c(e) = \chi - f(e)$ $\forall \in e\, E$.

Definition 1.1.32[32]

For two central soft sets (f,A) *and* (g,B) over a common universe χ , $(f,A) - (g,B)$ is defined to be a   central soft set $(f,A) \cap (g,B)^c$.

In   fact $(f,A) \sqcap (g,B)^c = (h,c)$, $C = A - B$ *and* ,

$$h_C(e) = \begin{cases} g^c(e) & \text{if } e \in A \cap B \\ f(e) & \text{if } e \in (B \cup A)^C \\ f(e) \cap g^c(e) & \text{other wise} \end{cases}$$

Theorem 1.1.33[32]

Let (f,A) *and* (g,B) be two central soft sets over a common universe χ , then

i) $[(f,A) \sqcap (g,B)]^C = (f,A)^C \sqcup (g,B)^c$

ii) $[(f,A) \sqcup (g,B)]^C = (f,A)^C \sqcap (g,B)^c$.

Theorem 1.1.34 [32]

Let (f,A) and (g,B) and (h,C) be a   central soft sets over a common universe χ , then ,

i) $(f,A) \sqcap [(g,B) \sqcap (h,C)] = [(f,A) \sqcap (g,B)] \sqcap (h,C)$

ii) $(f,A) \sqcup [(g,B) \sqcup (h,C)] = (f,A) \sqcup (g,B)] \sqcup (h,C)$

Theorem 1.1.35 [32]

Let (f,A) and (g,B) and (h,C) be a central soft sets over a common universe χ, then,

i) $(f,A) \sqcap [(g,B) \sqcup (h,C)] = [(f,A) \sqcap (g,B)] \sqcup [(f,A) \sqcap (h,C)]$

ii) $(f,A) \sqcup [(g,B) \sqcap (h,C)] = [(f,A) \sqcup (g,B)] \sqcap [(f,A) \sqcup (h,C)]$.

Definition 1.1.36[32]

Let (f,A) and (g,B) and (h,C) be central soft sets over a common universe χ, then,

$$(f,A) - (g,B) - (h,C) = (f,A) - [(g,B) \sqcup (h,C)] \ .$$

Theorem 1.1.37 [32]

Let $\{ (f_i, A_i), i \in I \}$ be a family of soft sets over a common universe χ, then, $A \subseteq \cup A_i$.

$\sqcup_{i \in I} (f_i, A_i) = (h, \sqcup_{i \in I} A_i)$.

Theorem 1.1.38 [32]

Let $\{ (f_i, A_i), i \in I \}$ be a family of soft sets over a common universe χ, then,

$(f,A) \sqcap (\sqcup_{i \in I} (f_i, A_i)) = \sqcup_{i \in I} \{ (f,A) \sqcap (f_i, A_i) \}$.

Fourth family

We will denote $S_4(\chi)$, and it contains the following type :

$$S_4(\chi) = \left\{ F_{i_A}, where \ A \subseteq E \ and \ F_i : A \longrightarrow IP(\chi) \right\}$$

The following definitions and concepts are so necessary to know how to deal with the properties of this type , as follows :

Definition 1.1.49 [33]

The soft union of the soft sets F_A *and* G_A *in* $S_4(\chi)$ over the same universe χ is a soft set H_A , such that :

$$H(e) = F(e) \cup G(e) \quad , \quad \forall\, e \in A.$$

We write $F_A \,\widetilde{\cup}\, G_A = H_A$.

Definition 1.1.40

For any two soft sets F_A and G_A , $a \in A$, the a −soft union and the a −soft intersection of F_A and G_A is defined as follows :

$$F_A \,\widetilde{\cup}_a\, G_A = H_a \text{ where } H_a = \{(a, F(a) \cup G(a))\}$$

$$F_A \,\widetilde{\cap}_a\, G_A = H_a \text{ where } H_a = \{(a, F(a) \cap G(a))\}$$

Let Λ be an arbitrary index set , then for any $a \in A$, the a −soft union and the a −soft intersection of the subfamily $\{F_{\lambda_A} , \lambda \in \Lambda\}$ of $S_A(\chi)$ is defined as follows :

$$\bigcup_{\lambda \in \Lambda} F_{\lambda_A} = H_a \text{ where } H_a = \{(a, \bigcup_{\lambda \in \Lambda} F_\lambda(a))\}$$

$$\bigcap_{\lambda \in \Lambda} F_{\lambda_A} = H_a \text{ where } H_a = \{(a, \bigcap_{\lambda \in \Lambda} F_\lambda(a))\}$$

Definition 1.1.41 [31]

The soft intersection of the soft sets F_A *and* G_A *in* $S_4(\chi)$ over the same universe χ is a soft set H_A , such that :

$$H(e) = F(e) \cap G(e) \quad , \quad \forall\, e \in A.$$

We write $F_A \,\widetilde{\cap}\, G_A = H_A$. F_A and G_A are said to be soft disjoint if $F_A \,\widetilde{\cap}\, G_A = \widetilde{\Phi}_A$.

Definition 1.1.42 [33]

For two soft sets F_A and G_A *in* $S_4(\chi)$_F_A is a soft subset of G_A and denoted by $F_A \,\widetilde{\subseteq}\, G_A$ iff $F(a) \subseteq G(a)$ $\forall\, a \in A$.

Definition 1.1.43 [31]

Two soft sets F_A and G_A *in* $S_4(\chi)$ are said to be equal if F_A is a soft subset of G_A and G_A is a soft subset of F_A .

Proposition 1.1.44 [33]

For any soft set F_A and $x \in \chi$, we say that $x \in_a F_A$ if $x \in F(a)$ $\forall\, a \in A$.

proposition 1.1.45 [31]

If F_A and G_A are two soft sets in $S_4(\chi)$, then ,

i) $[F_A \,\tilde{\cup}\, G_A]^C = F_A{}^C \,\tilde{\cap}\, G_A{}^C$

ii) $[F_A \,\tilde{\cap}\, G_A]^C = F_A{}^C \,\tilde{\cup}\, G_A{}^C$

Definition 1.1.46 [31]

Let I be an arbitrary index set and $\{\, F_{i_A}, i \in I\}$ be a sub family of $S_4(\chi)$,

The union of these soft sets is the soft set H_A where $H(a) = \cup i \in I\, F_i(a)$ for each $a \in A$,

we write $\cup i \in I\, F_{i_A} = H_A$

Example 1.1.47

Let $\chi = N$, $A = \{0,1\}$ and $I = \{\,1,2,3\dots\}$ for every $i \in I$ we consider the soft sets F_{i_A}, where the map $F_i : A \longrightarrow IP(\chi)$ it is defined as follows :

$$F_i(p) = \begin{cases} \{0,1,2,\dots 100\} & if\ p = 0 \\[2mm] \{100,101,102,..\} & if\ p = 1 \end{cases}$$

Then $\cup\{F_{i_A}, i \in I\,\} = F_A$, where the map $F : A \longrightarrow IP(\chi)$ defined as follows :

$$F(p) = \begin{cases} \{0,1,2,\dots 100\} & if\ p = 0 \\[2mm] \{100,101,102,..\} & if\ p = 1 \end{cases}$$

The intersection of the soft sets is the soft set M_A where $M(e) = \cap_{i \in I} F_i(e)$ for each $e \in A$, we write $\bigcap i \in I \; F_{i\,A} = M_A$.

Definition 1.1.48[31]

The complement of a soft set F_A denoted by $F_A{}^c$ is defined as :

$$(F_A)^c = F_A{}^c = \tilde{\chi}_A - F_A = \chi - F(a) \quad a \in A ,$$

F^c is called the soft complement function of F , clearly$(F^c)^c$ is the same as F , and $(F_A{}^c)^c = F_A$.

Definition 1.1.49[31]

A soft set F_A over χ is said to be the null soft set , denoted by $\widetilde{\Phi}_A$ if $\forall\, a \in A , F(a) = \varphi$.

Definition 1.1.50 [31]

A soft set F_A over χ is said to be the absolute soft set and denoted by $\tilde{\chi}_A$, if $\forall\, a \in A \; F(a) = \chi$.

Clearly $\tilde{\chi}_A - \tilde{\chi}_A = \widetilde{\Phi}_A$ and $\widetilde{\Phi}_A{}^c = \tilde{\chi}_A$.

Definition 1.1.51

Let F_A be any soft set over the universe χ , for $a \in A$ we say that a soft point $x_a \,\widetilde{\in}\, F_A , x \in \chi$ is an a −empty soft point and denoted by $\widetilde{\Phi}_a$ is defined as follows :

$$\widetilde{\Phi}_a = \{\, (a, \varphi) , \}$$

And we say that x_a is an a −absolute soft point and denoted by $\tilde{\chi}_a$, is defined as follows :

$$\tilde{\chi}_a = \{\, (a, \chi) \,\} \quad , \text{note that} \quad \tilde{\chi}_A = \{\, (a, \chi) , \forall a \in A \}.$$

Definition1.1.52

Let F_A be any soft set over χ, and let $a \in A$,we say that a point $x \in \chi$ belong to the soft set F_A and denoted by $x_a \,\in_a\, F_A$, that is $x \in F(a)$.

Also we say that a soft set G_A is a soft subset of F_A [at the point x] and denoted by : $G_A \,\widetilde{\subseteq}_a\, F_A$, if $G(a) \subseteq F(a)$.

The principle that has been reliable to define the concepts of a soft point is by $(A \times \chi)$ which is defined on $A \times IP(\chi)$.

Where , for $a \in A$ we have $\{a\} \times F(a) \subseteq A \times IP(\chi)$, it gives an ordered pairs as $(a, x), x \in F(a)$ and :

$$\bigcup_{x \in F(a)}(a, x) = \left(a, F(a)\right)$$

And the union of these ordered pairs $\left(a, F(a)\right)$ gives the original soft set F_A that is :

$$F_A = \bigcup_{a \in A}\left\{\left(a, F(a)\right), a \in A \right\}.$$

proposition 1.1.53 [31]

i) $\{\bigsqcup_{i \in I} [F_{i_A}]\}^c = \bigsqcap_{i \in I} [F_{i_A}]^c$

ii) $\{\bigsqcap_{i \in I} [F_{i_A}]\}^c = \bigsqcup_{i \in I} [F_{i_A}]^c.$

Example 1.1.54

Let t $\chi = R$, $A = \{0, 1, 2\}$ and we consider the soft set F_A where the map $F: A \longrightarrow IP(\chi)$ is defined as follows :

$F(p) = [p, 0)$ for every $p \in A$, then $\tilde{\chi}_A - F_A = H_A$ where the map $H: A \longrightarrow IP(\chi)$ is defined as follows :

$H(p) = [0, p)$ for every $p \in A$.

Both of the definitions of union and intersections that are defined on $S_4(\chi)$ are equivalent with all the definitions that are defined on $S_2(\chi)$ and $S_1(\chi)$.

That is , there is no problem results on these definitions when we are using this type of a soft sets on the other definitions .

proposition 1.1.55[31]

If F_A and G_A are two soft sets in $S_4(\chi)$, then , the following statements are true :

i) $F_A \tilde{\cap} \tilde{\chi}_A = F_A$

ii) $F_A \,\widetilde{\cup}\, \widetilde{\Phi}_A = F_A$,

iii) $F_A \,\widetilde{\cap}\, \widetilde{\Phi}_A = \widetilde{\Phi}_A$

iv) $F_A \,\widetilde{\cup}\, \tilde{\chi}_A = \tilde{\chi}_A$

proposition 1.1.56 [31]

If F_A and G_A are two soft sets in $S_4(\chi)$, then , the following statements are true :

i) $F_A \,\widetilde{\subseteq}\, G_A$ iff $F_A \,\widetilde{\cap}\, G_A = F_A$

ii) $F_A \,\widetilde{\subseteq}\, G_A$ iff $F_A \,\widetilde{\cup}\, G_A = G_A$.

proposition 1.1.57 [31]

If F_A *and* G_A are two soft sets in $S_4(\chi)$, then , the following statements are true :

1) If $F_A \,\widetilde{\cap}\, G_A = \widetilde{\Phi}_A$, then $F_A \,\widetilde{\subseteq}\, G_A{}^c$

2) $F_A \,\widetilde{\cup}\, F_A{}^c = \tilde{\chi}_A$

3) If $F_A \,\widetilde{\subseteq}\, G_A$ and $G_A \,\widetilde{\subseteq}\, H_A$ then $F_A \,\widetilde{\subseteq}\, H_A$

4) If $F_A \,\widetilde{\subseteq}\, G_A$ and $H_A \,\widetilde{\subseteq}\, S_A$, then $F_A \,\widetilde{\cap}\, H_A \,\widetilde{\subseteq}\, G_A \,\widetilde{\cap}\, S_A$

5) $F_A \,\widetilde{\subseteq}\, G_A$ iff $G_A{}^c \,\widetilde{\subseteq}\, F_A{}^c$

Note 1.1.58 [33]

$\widetilde{\Phi}_A \,\widetilde{\subseteq}\, F_A \,\widetilde{\subseteq}\, \tilde{\chi}_A$.

proposition1.1.59 [37]

If F_A and G_A are two soft sets in $S_4(\chi)$, then , the following statements are true :

i) $F_A \,\widetilde{\cap}\,(G_A \,\widetilde{\cap}\, H_A) = (F_A \,\widetilde{\cap}\, G_A)\,\widetilde{\cap}\, H_A$

ii) $F_A \,\widetilde{\cup}\,(G_A \,\widetilde{\cup}\, H_A) = (F_A \,\widetilde{\cup}\, G_A)\,\widetilde{\cup}\, H_A$

iii) $F_A \,\widetilde{\cap}\,(G_A \,\widetilde{\cup}\, H_A) = (F_A \,\widetilde{\cap}\, G_A)\,\widetilde{\cup}\,(F_A \,\widetilde{\cap}\, H_A)$

iv) $F_A \,\widetilde{\cup}\,(G_A \,\widetilde{\cap}\, H_A) = (F_A \,\widetilde{\cup}\, G_A)\,\widetilde{\cap}\,(F_A \,\widetilde{\cup}\, H_A)$.

proposition 1.1.60 [37]

Let I be an arbitrary index set , and let $\{\, F_{i_A}\,, i \in I\}$ be a sub family of $S_4(\chi)$ then , the following statements are true :

i) $F_{i_A} \widetilde{\subseteq} \widetilde{U}\{F_{i_A} \ i \in I\}$ for every $i \in I$

ii) $\widetilde{\cap}\{F_{i_A} \ i \in I\} \widetilde{\subseteq} F_{i_A}$ for every $i \in I$.

Definition 1.1.61 [37]

Let F_A and G_A be two soft sets in $S_4(\chi)$, then the soft symmetric deference between F_A and G_A is the soft set $H_A \in S_4(\chi)$, where the map $H: A \longrightarrow IP(\chi)$ is defined as follows :

$H(p) = \left(F(p)\backslash G(p) \right) \cup \left(G(p)\backslash F(p) \right)$ for every $p \in A$, symbolically we write $H_A = F_A \ \triangle \ G_A$.

Example1.1.62

Let $\chi = \{1,2,3,4,5,6,7,8\}$ and $A = \{0,1,2,3 \dots\}$ we consider the soft sets let $F_A \ and \ G_A$ where the maps $F: A \longrightarrow IP(\chi)$ and $G: A \longrightarrow IP(\chi)$, is defined as follows :

$$F(p) = \begin{cases} \{1,2,3\} & if \ p = 0 \\ \varphi & other \ wise \end{cases}$$

$$G(p) = \begin{cases} \{1,4,5\} & if \ p = 0 \\ \varphi & other \ wise \end{cases}$$

Then $\ F_A \ \triangle \ G_A = H_A$ where the map $H: A \longrightarrow IP(\chi)$ is defined as follows

$$H(p) = \begin{cases} \{2,3,4,5\} & if \ p = 0 \\ \varphi & other \ wise \end{cases}$$

Conclusion

1. Through the mentioned properties of each of the four families , we noted that there are many relations between them , all of $S_2(\chi), S_3(\chi)$ and $S_4(\chi)$ are special cases of $S_1(\chi)$ and there are several properties of the soft union and the soft intersection is true in a family and false in the other , such as :
   (a)   In [32] , the soft intersection cannot be true from a topological view but it is true in $S_4(\chi)$ .as the following :

      If $\widetilde{}$ $\widetilde{\chi}_A$, $F_A \widetilde{\in} S_4(\chi)$,   then

$$H_A(a) = \begin{cases} \tilde{\chi}_A(a) = \varphi & \text{if } a \in A \setminus A = \varphi \\ F(a) = \varphi & \text{if } a \in A \setminus A = \varphi \\ \tilde{\chi}_A(a) \cap F(a) = \varphi & \text{other wise} \end{cases}$$

$= \widetilde{\Phi}_A$. this property does not hold in the other families .

(b)　All the properties of the soft union and the soft intersection in $S_1(\chi), S_2(\chi)$ and $S_3(\chi)$ are true in $S_4(\chi)$. That is $S_4(\chi)$ comprehensive considered the family to bring all the properties .

2.　Previously, we understand that each of this four family have a properties different from the other . The first family has been defined a soft set by changing both of the mapping F and the sets of parameters , the second family has been defined as a soft set by putting the mapping F and changing the sets of parameters at each time , but in the third family the problem is different , because of existing a fundamental set containing of the parameters that is E_χ , and this set has defined all the soft sets . In the fourth family , they are defined as soft set by a changing the mapping F and putting the set of parameters .

It is clear that in each type of them , there are many definitions and properties different in each one of them like , soft intersection , soft union , soft complement , soft subset , absolute soft set and null soft set . The relationship between these concepts are studied in [1,2,4,10,13] and several operations , like the distribution lows and De-Morgan low's and another relations .

From above , we will depend on the fourth family in this study , and we will use the same mathematical dynamic which is used in the other families .

From a topological opinion , the union and the intersection forms the foundation that is reliable in deferent subjects , consequently , we will introducing An analytical study for these definitions from a topological point of view, and we will depend on the important conditions that must be done in general topological spaces to know the power of each definition or the negation of it , as follows :

<u>Note1.1.63</u>

In the following , An analytical study of the most important topological operations which will form the basis of the soft topological spaces:

In the following discussion ,we will take a look and discussing several definitions that are defined by researches for the concepts of soft union and the soft intersection of a soft sets . and then we will test their propriety in the soft topology .

We will depend here on two conditions to know which of the definitions is true :

For a soft set F_A over the universe χ , then :

$$(1) \quad F_A \,\tilde{\cup}\, \tilde{\chi}_A = \tilde{\chi}_A$$
$$(2) \quad F_A \,\tilde{\cap}\, \tilde{\Phi}_A = \tilde{\Phi}_A$$

This is because of the first condition is necessary in the soft topology , if it is not , then we cannot guarantee the emergence of the absolute soft set $\tilde{\chi}_A$, and similarly for the second condition then we cannot guarantee the emergence of the null soft set $\tilde{\Phi}_A$ .We are know one of the two sets $\tilde{\chi}_A$ and $\tilde{\Phi}_A$ are disappear , then the space cannot be a soft topological space (soft subspace) , we will also show later .

In the following table we include all the definitions of the soft union and the soft intersection that are defined , to know which of them implies to the truing of a soft topology or not .

Soft Union \ Soft Intersection	-1-	-2-	-3-	-4-
	N.Cageman ,S.Karatas , [4]2011	M.ifran Ali , F.feng ,X. liu W.Keun min and M. shabir [11] 2009	F.Feng, Y.B.JUN,and X.Z.Zhao [40] 2011	W.K.Min[25] 2011
1 N.Cageman ,S.Karatas , [4]2011	√	×	√	√

2	M.ifran Ali , F.feng ,X. liu W.Keun min and M. shabir [11] 2009	√	×	√	√
3	F.Feng, Y.B.JUN,and X.Z.Zhao [40] 2011	×	×	×	×
4	W.K.Min[25] 2011	√	×	√	√

The error in definition (2,2) is in the concept of the soft intersection , since it does not achieve the second condition , and this implies an error of existing the empty set in a soft topology . Thus , this pair cannot be true .

The definitions (1,2) , (3,1) , (3,2) , (3,4) ,(4,2) are also incorrect , because there is an error in one of the two conditions above , hence it cannot be true in a soft topology .

But the definitions (1,1) ,(1,3) , (1,4) , (2,1) , (2,3) , (2,4) ,(4,1) ,(4,3) ,(4,4) are true , and there is no error in using the soft topology .

In the following , Some of basic concepts and illustrative examples in soft set theory by using the fourth family $S_4(\chi)$.

Definition 1.1.64[31]

The soft set F_A over the universe χ is said to be soft point and denoted by a_F if for the element $a \in A$ then $F(a) \neq \varphi$

and $F(p) = \varphi$ for all $p \in A \setminus \{a\}$.

On the other hand $a_F(a) = \{(a, F(a))\} \cup \{(a, \varphi); a \in A\}$ or simply we write $a_F = \{(a, F(a))\}$ and $\tilde{\chi}_A - a_F = \begin{cases} \chi - F(a) & if\ p = a \\ \chi & if\ p \neq a \end{cases}$

Definition 1.1.65

The soft set F_A over the universe χ is called single soft point if $F(a) = \{x\} \quad \forall\, a \in A$, we will denote by

x_e , such that $x_e = \{\, \{\,(a, x)\}\,;\, a \in A\}$, that is $\tilde{\chi}_A - x_e = \chi - \{x\}\, \forall a \in A$

Definition 1.1.66

For $a \in A$ and $x \in \chi$ we say that the soft set x_a in $S_4(\chi)$ is $a-$ soft point if

$$x_a(p) = \begin{cases} \{x\} & if\ p = a \\ \varphi & if\ p \neq a \end{cases}$$

That is $x_a(p) = \{(a, x)\} \cup \{(a, \varphi), p \in A\}$ or simply we denote by $x_a(p) = \{(a, \{x\})\}$, that is

$$\tilde{\chi}_A - x_a(p) = \begin{cases} \chi - \{x\} & if\ p = a \\ \chi & if\ p \neq a \end{cases}$$

Example 1.1.67

Suppose that $\chi = \{x_1, x_2, x_3\}$ and $A \subseteq E$ such that $A = \{a_1, a_2\}$.

Let $F_A = \{(a_1, \{x_1, x_2\}), (a_2, \varphi)\} = e_F$ is a soft point .

$G_A = \{(a_1, \{x_1\}, (a_2, \{x_1\}))\} = x_e$ is a single soft point .

$x_{1_{a_1}} = \{(a_1, \{x_1\}(a_2, \varphi)\}$. is an $a_1 -$soft point .

Remark 1.1.68

Note that a_F , x_e and x_a are different in definition $[1.1.61\,,\,1.1.62\,,\,1.1.63]$.

Definition 1.1.69[31]

1. A soft point a_F is said to be in the soft set G_A , and denoted by

$a_F \,\tilde{\subseteq}\, G_A$, if for the element $a \in A$, $F(a) \,\tilde{\subseteq}\, G(a)$.

2.   A single soft point x_e satisfying the following property :

$x_e \,\tilde{\in}\, G_A$ iff $x \in G(a)\ \forall\, a \in A$

Remark 1.1.70

An element $\alpha \in F_A$ is called a singleton element if :

1. $\alpha = \big(a, F(a)\big)$ for some $a \in A$.
2. $F(a)$ is a singleton set (i.e $F(a) = \{x\}$ for $x \in \chi$

proposition 1.1.71

let F_A be a soft set over the universe χ , let e_G be any soft point over χ

, if $a_G \,\widetilde{\in}\, F_A$, then $F_A \,\widetilde{\cap}\, a_G \neq \widetilde{\Phi}_A$

Proof \\

Let $a_G \in F_A$, then there is a $\in A$ such that $G(a) \neq \varphi$ and $G(a) \subseteq F(a)$

This means that $G(a) \cap F(a) \neq \varphi$ other wise $G(a) \cap F(p) = \varphi$ $\forall$ $a \neq p$

Thus $F_A \,\widetilde{\cap}\, a_G \neq \widetilde{\Phi}_A$.

The opposite case is incorrect . The following example shows this fact:

Example 1.1.72

Suppose that the universe χ containing of five houses ; $\chi = \{h_1, h_2, h_3, h_4, h_5\}$ and $A = \{a_1, , a_2, a_3\}$, and
$F_A = \{(a_1, \{h_1, h_3,\}), (a_2, \{h_3, h_4, h_5,\}), (a_3, \{h_2, h_3, h_5,\})\}$

and $G_A = \{(a_1, \{ h_1, h_3, h_4,\}), (a_2, \{\varphi\}), (a_3, \{\varphi\})\}$ note that G_A is a soft point , i.e $a_G = \{(a_1, \{ h_1, h_3, h_4,\})\}$ and $F_A \,\widetilde{\cap}\, a_G \neq \widetilde{\Phi}_A$.

proposition 1.1.73 [37]

for any soft set F_A and a single soft point x_e over χ , then $x_e \,\widetilde{\subseteq}\, F_A$ iff $F_A \,\widetilde{\cap}\, x_e \neq \widetilde{\Phi}_A$.

(1.2) Soft functions:

One of the important concepts in soft set theory is knowing what is the structure of a soft functions .This section is highlighted on the logical sequence to the mechanism which is formed by these functions . In addition to the basic definitions and the properties of this object .

Definition1.2.1[42]

Let χ and Y be two initial universe sets and A, B be sets of parameters , $u: \chi \longrightarrow Y$ and $p: A \longrightarrow B$ be a mapping , then the mapping :

$f: S_1(\chi) \longrightarrow S_1(Y)$ on A and B respectively is denoted by f_{pu} and defined as :for a soft set F_A in $S_1(\chi)$, then $\left(f_{pu}(F_A), B\right)$, $B = p\,(A) \subseteq B$ is a soft set in $S_1(\,Y\,)$ given by:

$$f_{pu}(F_A)(\beta) = \begin{cases} u\left(\displaystyle\bigcup_{\alpha \in p^{-1}(\beta) \cap A} (F(\alpha))\right), & if\, p^{-1}\,(\beta) \cap A \neq \varphi \\ \varphi & other\ wise \end{cases}$$

For $\beta \in B$, $f_{pu}(F_A)(\beta)$ is called a soft image of F_A .

Definition 1.2.2[42]

Let $f: S_1(\chi) \longrightarrow S_1(Y)$ be a soft mapping from the soft family $S_1(\chi)$ into the soft family $S_1(Y)$ and let G_B be a soft set over Y . Let $u: \chi \longrightarrow Y$ and $p : A \longrightarrow B$ be mappings , then $\left(f_{pu}^{\,-1}(G_B), C\right)$, $C = p^{-1}(\,B\,)$ is a soft set in $S_1(\chi)$ defined as :

$$f_{pu}^{\,-1}(G_B)(\alpha) = \begin{cases} u^{-1}(G(p(\alpha)), & if\ p(\alpha) \in B \\ \varphi & other\ wise \end{cases}$$

For $\alpha \in A$, $f_{pu}^{\,-1}(G_B)(\alpha)$ is called the inverse image of G_B .

We will introduce a new definition of a soft mappings .   Note :

Definition1.2.3

Let χ and Y be two initial universal sets and A, B be sets of parameters , $u: \chi \longrightarrow Y$ and $p: A \longrightarrow B$, then the mapping :

$f: (\chi, A) \longrightarrow (Y, A)$ (I .e $f: S_4(\chi) \longrightarrow S_4(Y)$) on A and B respectively is denoted by f_{pu} and can be shown as:

$$f_{pu} = \left\{ \left(f_{pu}(F_A), p(A) \right), p(A) \subseteq B \right\}.$$

Where

$f_{pu}(F_A)(\beta) =$

$$\begin{cases} u\left(\bigcup_{\alpha \in p^{-1}(\beta) \cap A \neq \varphi}(F(\alpha)) \right), & if \ \ p^{-1}(\beta) \neq \varphi \\ \quad \varphi & other\ wise \end{cases}$$

For $\beta \in B \ \exists \ a \in p(A) \ such \ that \ p(a) = \beta$. that is $p^{-1}(\beta) \neq \varphi$

Since $p^{-1}(\beta) \subseteq A$, hence $p^{-1}(\beta) \cap A \neq p^{-1}(\beta)$, hence we get that

$$f_{pu}(F_A)(\beta) = u\left(\bigcup_{\alpha \in p^{-1}(\beta)} F(\alpha) \right)$$

Constructing :

$\forall A \neq \varphi$ and since p is a mapping , so $p(A) \neq \varphi$, that is $\forall$ $\beta \in p(A) \exists \ a \in A \ such \ that \ p(a) = \beta$ $and \ p^{-1}(\beta) \neq \varphi$ since $a \in p^{-1}(p(a))$ so :

$f_{pu}(F_A)(\beta) = u\{ \bigcup_{\alpha \in p^{-1}(\beta)}(F(\alpha)) \} \forall \beta \in p(A)$.

- if p is a one to one (1-1) , then $p^{-1}(p(A)) = A$, that is $\forall$ $\beta \in p(A) \exists \ a \in A \ such \ that \ p(a) = \beta$ [note : $p^{-1}(\beta) = p^{-1}(p(a)) = \{a\}$] $and \ f_{pu}(F_A)(\beta) = u(F(a))$.

now Let χ and Y be two initial universal sets and A, B be a sets of parameters , $u: \chi \longrightarrow Y$ and $p: A \longrightarrow B$, then the inverse image of the soft mapping :

$f: (\chi, A) \longrightarrow (Y, A)$ is denoted by f_{pu}^{-1} and defined as :

for any $G_B \in S_1(Y)$, then

$$f_{pu}^{-1}(G_B)(\alpha) = \begin{cases} u^{-1}(G(p(\alpha))), & if \ p(\alpha) \in B \\ \varphi & other\ wise \end{cases}$$

Note that $\left(f_{pu}^{-1}(G_B), p^{-1}(\beta)\right) = \left(f_{pu}^{-1}(G_B), A\right)$ because p is a mapping so , $p^{-1}(\beta) = A$

$\forall\, \alpha \in p^{-1}(\beta) = A$ and G_B is a soft set over Y , $\left(f_{pu}^{-1}(G_B), A\right)$ is called a soft inverse image of G_B , simply we write $\left(f_{pu}^{-1}(G_B), A\right)$ as $f_{pu}^{-1}(G_B)$.

Remark 1.2.4

For each $a \in A$ and $x \in \chi$, then we can define the soft mapping f_{pu} on $a - $ soft point x_a , as follows :

1. $f_{pu}(x_a)(\beta) = u\left(\bigcup_{\alpha \in p^{-1}(\beta)} x_a(\alpha)\right) =$
$$\begin{cases} u(x) & if\ \alpha \in p^{-1}(\beta) \\ \varphi & if\ \alpha \notin p^{-1}(\beta) \end{cases}$$

(I .e $\left(f_{pu}(x_a)\right)(p(a)) = p(a)$

I .e $\left(\left(f_{pu}(x_a)\right)(p(a)) = \{(p(a), \{x(a)\})\}\right.$

2. Now , for $b \in B$ and $y \in Y$,
$$f_{pu}^{-1}(y_b)(a) = \begin{cases} u_b^{-1}(y) & if\ b = p(a) \\ \varphi & if\ b \neq p(a) \end{cases}.$$

$$= u_b^{-1}\left(y_b(p(a))\right)$$

proposition 1.2.5 [37]

Let $F_A, F_{1_A} \in S_1(\chi)$ and $G_B, G_{1_B} \in S_1(Y)$, the following statements are true :

i) if $F_A \widetilde{\subseteq} F_{1_A}$, then $f_{pu}(F_A) \widetilde{\subseteq} f_{pu}(F_{1_A})$.

ii) $G_B \widetilde{\subseteq} G_{1_B}$, then $f_{pu}^{-1}(G_B) \widetilde{\subseteq} f_{pu}^{-1}(G_{1_B})$

iii) $F_A \widetilde{\subseteq} f_{pu}^{-1} f_{pu}(F_A))$

iv) if p is a (1-1) map of χ into Y and u is a (1-1) of A into B then ,
   $F_A = f_{pu}^{-1}(f_{pu}(F_A))$

v) $f_{pu}\left(f_{pu}^{-1}(G_B)\right) \widetilde{\subseteq} G_B$

vi) If p is a map from χ into Y and u is a map of A into B then ,

$$f_{pu}\left(f_{pu}{}^{-1}(G_B)\right) = G_B$$

vii) $f_{pu}{}^{-1}(\tilde{\chi}_A - G_B) = \tilde{\chi}_A - f_{pu}{}^{-1}(G_B).$

Remark 1.2.6

For any $x_1, x_2 \in \chi$ and $y_1, y_2 \in Y$ and $a \in A$, $b \in B$, the flowing statements are true :

(1) $x_{1_a} \tilde{\subseteq} x_{2_a}$, $f_{pu}(x_{1_a}) \tilde{\subseteq} f_{pu}(x_{2_a})$

(2) $y_{1_b} \tilde{\subseteq} y_{2_b}$, $f_{pu}{}^{-1}(y_{1_b}) \tilde{\subseteq} f_{pu}{}^{-1}(y_{2_b})$

(3) $x_{1_a} \tilde{\subseteq} f_{pu}{}^{-1}(x_{1_a})$

(4) $f_{pu}\left(f_{pu}{}^{-1}(y_{1_b})\right) \tilde{\subseteq} y_{1_b}$

(5) If p is a map from χ into Y , and u is a map of A into B , then :

$$f_{pu}\left(f_{pu}{}^{-1}(y_{1_b})\right) = y_{1_b} \; .$$

(6) $\quad f_{pu}{}^{-1}(\tilde{\chi}_b - y_{1_b}) = \tilde{\chi}_a - f_{pu}{}^{-1}(y_{1_b}).$

proposition 1.2.7 [37]

Let I be an arbitrary index set , $\{\, F_{i_A}, i \in I \,\} \tilde{\subseteq} S_4(\chi)$ and $\{G_{i_B}, i \in I \,\} \tilde{\subseteq} S_4(Y)$, the following statements are   true :

i) $\quad f_{pu}(\tilde{\cup}\{\, F_{i_A}, i \in I \,\} = \tilde{\cup} \{ f_{pu}(F_{i_A}), i \in I \,\}$

ii) $\quad f_{pu}(\tilde{\cap}\{\, F_{i_A}, i \in I \,\} \tilde{\subseteq} \tilde{\cap} \{ f_{pu}(F_{i_A}), i \in I \,\}$

iii) $\quad f_{pu}{}^{-1}(\tilde{\cup}\{\, G_{i_B}, i \in I \,\} = \tilde{\cup} \{ f_{pu}{}^{-1}(G_{i_B}), i \in I \,\}$

iv) $\quad f_{pu}{}^{-1}(\tilde{\cap}\{\, G_{i_B}, i \in I \,\} = \tilde{\cap} \{ f_{pu}{}^{-1}(G_{i_B}), i \in I \,\}.$

The same proposition is   true if we deal with a point $a \in A$.

Chapter Two

Soft Topology

Introduction

The soft set theory included many previous studies by several researchers that highlight the problems which are needed to be solved in this theory , among these problems is the soft topology .Here , this chapter is divided into two sections ,the first section contains the soft topology , and second contains of many soft topological concepts which explained the points in the soft topology .

(2.1) Soft topology

This section studied the soft topology and the soft subspace in addition to the soft continuous mappings .

Definition 2.1.1 [31]

Let χ be an initial universal set and A be a set of parameters , and $\tilde{\tau} \cong S_A(\chi)$, we say that the family $\tilde{\tau}$ is a soft topology on χ if the following axioms hold:

$[T_1]$: $\tilde{\Phi}_A , \tilde{\chi}_A \,\tilde{\in}\, \tilde{\tau}$

$[T_2]$: if G_A and $H_A \,\tilde{\in}\, \tilde{\tau}$, then $G_A \,\tilde{\cap}\, H_A \,\tilde{\in}\, \tilde{\tau}$.

$[T_3]$: if $G_{i_A} \,\tilde{\in}\, \tilde{\tau}$ for every $i \in I$, then $\tilde{\cup} \,\{G_{i_A}, i \in I\} \,\tilde{\in}\, \tilde{\tau}$.

The triple $(\tilde{\chi}_A, \tilde{\tau}, A)$ is called soft topological space or soft space .

❖ The members of $\tilde{\tau}$ are called soft open sets .
❖ A soft set F_A is called soft closed set if the complement $(\tilde{\chi}_A - F_A)$ is soft open set(belong to $\tilde{\tau}$) .
❖ The family of all soft closed sets are denoted by $C(\tilde{\chi})$ and defined as follows :

$$C(\tilde{\chi}_A) = \{\tilde{\chi}_A - F_A , F_A \,\tilde{\in}\, \tilde{\tau}\} .$$

Proposition 2.1.2[38]

Let $(\tilde{\chi}_A, \tilde{\tau}, A)$ be a soft topological space , the family $C(\tilde{\chi})$ has the following properties :

(1) $\tilde{\Phi}_A , \tilde{\chi}_A \,\widetilde{\in}\, C(\tilde{\chi}_A)$

(2)   if G_A and $H_A \,\widetilde{\in}\, C(\tilde{\chi}_A)$, then $G_A \,\widetilde{\cup}\, H_A \,\widetilde{\in}\, C(\tilde{\chi}_A)$

(3)   if $G_{i_A} \,\widetilde{\in}\, C(\tilde{\chi}_A)$ for every $i \in I$, then $\widetilde{\cap} \{G_{i_A}, i \in I\} \,\widetilde{\in}\, C(\tilde{\chi})$.

Remark 2.1.3 [27]

See the examples [4.4 ,4.5, 4.6 ,4.7] .

Definition 2.1.4 [31]

Let $(\tilde{\chi}_A, \tilde{\tau}, A)$ be a soft topological space , a soft set G_A is called a soft neighborhood (briefly :nhd) of a soft point a_F over χ , if there exists a soft open set H_A such that $a_F \,\widetilde{\in}\, H_A \,\widetilde{\subseteq}\, G_A$.

The neighborhoods system of a soft point a_F denoted by $N_{\tilde{\tau}}(a_F)$ is the family of all its neighborhoods.

Theorem 2.1.5[31]

Let $(\tilde{\chi}_A, \tilde{\tau}, A)$ be a soft topological space , and let G_A be a soft set over χ , then :

A soft set G_A over χ is a soft open set iff it is a soft nhd at each of its elements , [this is true for any point $a \in A$].

Theorem 2.1.6 [31]

Let $(\tilde{\chi}_A, \tilde{\tau}, A)$ be a soft topological space, The neighborhood system $N_{\tilde{\tau}}(a_F)$ at a_F has the following properties :

(1)   if $G_A \in N_{\tilde{\tau}}(a_F)$ then $a_F \in G_A$

(2)   if $G_A \in N_{\tilde{\tau}}(a_F)$ and $G_A \,\widetilde{\subseteq}\, H_A$ then $H_A \in N_{\tilde{\tau}}(a_F)$.

(3)   if $G_A, H_A \in N_{\tilde{\tau}}(a_F)$, then $G_A \,\widetilde{\cap}\, H_A \in N_{\tilde{\tau}}(a_F)$.

(4)   if $G_A \in N_{\tilde{\tau}}(a_F)$ then there exists $M_A \in N_{\tilde{\tau}}(a_F)$ such that $G_A \in N_{\tilde{\tau}}(a^-{}_M)$ for each $a^-{}_M \,\widetilde{\in}\, M_A$.

Definition 2.1.7 [37]

Let $(\tilde{\chi}_A, \tilde{\tau}, A)$ be a soft topological space , $a \in A$, $x \in \chi$, we say that a soft set $F_A \tilde{\in} \tilde{\tau}$ is an a-soft open nhd of x in $(\tilde{\chi}_A, \tilde{\tau}, A)$ *if* $x \in F(a)$.

Definition 2.1.8

Let $(\tilde{\chi}_A, \tilde{\tau}, A)$ be a soft topological space , $a \in A$, $x \in \chi$, a soft set F_A is an a-soft nhd of $x \in \chi$, if there is an a-soft open set $G_{(a,x)}$ of x such that $G_{(a,x)} \tilde{\subseteq} F_A$.Where $G_{(a,x)}$ is a soft open set containing x ,

that is $[\, x \in G(a)\,]$.

Note2.1.9

Every a-soft open nhd of any point $x \in \chi$ is an a $-$soft nhd of x but the converse may not be true . as it is clear in the following example .

Example 2.1.10

Let χ be a universal set such that $\chi = \{x_1, x_2, x_3\}$ and let $A \subseteq E$ be a subset of a set of parameters such that $A = \{a_1, a_2, a_3\}$

$$\tilde{\tau} = \left\{\tilde{\Phi}_A, \tilde{\chi}_A, F_{1_A}, F_{2_A}\right\} \text{ be a soft topological space where :}$$

$$F_{1_A} = \{(a_1, \{x_1, x_2\}), (a_2, \{x_2\}), (a_3, \varphi)\}$$

$$F_{2_A} = \{(a_1, \{x_1\}), (a_2, \varphi), (a_3, \varphi)\}$$

Now , if we consider the soft set F_A as follows :

$F_A = \{(a_1, \{x_1, x_2\}), (\, a_2, \{x_1, x_2, x_3\}), (a_3, \varphi)\}$ be a soft set over χ , then :for a point $a_1 \in A$ and $x_1 \in \chi$, $F_{1_A} = \{(a_1, \{x_1, x_2\}), (a_2, \{x_2\}), (a_3, \varphi)\}$ is a soft open set containing x_1 and $F_{1_A} \tilde{\subseteq} F_A$, that is F_A is a_1 $-$soft nhd of x_1 but not a_1 $-$soft open nhd of x_1

Definition 2.1.11[31]

Let $(\tilde{\chi}_A, \tilde{\tau}, A)$ be a soft topological space , and let G_A be a soft set over the universe χ , then :

The soft interior of G_A is a soft open set defined as :

$$IntG_A = \tilde{\cup} \{S_A, S_A \text{ is soft open and } S_A \tilde{\subseteq} G_A \}$$

It is clear that $Int\ G_A$ a largest soft open set contained in G_A.

Definition 2.1.12 [31]

Let $(\tilde{\chi}_A, \tilde{\tau}, A)$ be a soft topological space , and let G_A be a soft set over χ then , the soft point $e_F \ \tilde{\in}\ \tilde{\chi}_A$ is called soft interior point of a soft set G_A if there exists a soft open set H_A such that $e_F \ \tilde{\in}\ H_A \tilde{\subseteq} G_A$.

Proposition2.1.13 [31]

Let $(\tilde{\chi}_A, \tilde{\tau}, A)$ be a soft topological space , and let G_A be a soft set over χ , then :

$$IntG_A = \tilde{\cup}_{e \in A} \{e_F, e_F \text{ is any soft interior point of } G_A, e \in A\}$$

Definition2.1.14[37]

Let $(\tilde{\chi}_A, \tilde{\tau}, A)$ be a soft topological space and $a \in A$, a point $x \in \chi$ is said to be a −interior point of a soft set F_A if there exists an a −soft open nhd $G_{(a,x)}$ of x such that $G_{(a,x)} \tilde{\subseteq} F_A$, the set of all a −soft interior points of F_A is denoted by $a - IntF_A$.

Example 2.1.15

Let χ be a universal set such that $\chi = \{x_1, x_2, x_3\}$ and let $A \subseteq E$ be a subset of a set of parameters such that $A = \{a_1, a_2, a_3\}$

$\tilde{\tau} = \{\tilde{\Phi}_A, \tilde{\chi}_A, F_{1_A}, F_{2_A}\}$ be a soft topological space such that :

$$F_{1_A} = \{(a_1, \{x_1, x_2\}), (a_2, \{x_2\}), (a_3, \varphi)\}$$

$$F_{2_A} = \{(a_1, \{x_1\}), (a_2, \{\varphi\}), (a_3, \varphi)\}$$

Now , if we consider the soft set F_A as follows :

$F_A = \{(a_1, \{x_1, x_2\}), (a_2, \{x_1, x_2, x_3\}), (a_3, \varphi)\}$ be a soft set over χ , then :

now for a point $a_1 \in A$, we have :

since $x_1 \in F(a_1)$ and $\exists F_{1_A} \in \tilde{\tau}$ such that $x_1 \in F_1(a_1) \subseteq F(a_1)$, hence by [Definition 2.1.14] x_1 is an $a_1 -$ interior point of F_A , similarly for $x_2 \in F(a_1), \exists F_{1_A} \in \tilde{\tau}$ such that $x_2 \in F_1(a_1) \subseteq F(a_1)$,

so x_2 is an $a_1 -$soft interior point $of \, F_A$, and x_3 is not

hence , $a_1 - Int(F_A) = \{(a_1, \{x_1, x_2\}), (a_2, \varphi), (a_3, \varphi)\}$, for a_2 we have :

$x_1 \in F(a_2)$ and $\exists F_{1_A} \in \tilde{\tau}$ such that $x_1 \in F_1(a_2) \subseteq F(a_2)$, thus x_1 is an $a_2 -$soft interior point $of \, F_A$, similarly $x_2 \, and \, x_3$ are not $a_2 -$soft interior point $of \, F_A$,

hence $a_1 - Int(F_A) = \{(a_2, \{x_2\}), (a_1, \varphi), (a_3, \varphi)\}$.

Proposition 2.1.16

Let $(\tilde{\chi}_A, \tilde{\tau}, A)$ be a soft topological space , and F_A be any soft set over χ , , for $a \in A$:

$$\bigcup_{a \in A}(a - IntF_A) = IntF_A$$

Proof//

Let $x \in \bigcup_{a \in A}(a - IntF_A)$, iff $\exists \, a \in A$,such that $x \in a - IntF_A$

Iff $\exists \, an \, a - soft \, open \, nhd \, G_{(a,x)}$ of x such that $x \in G_{(a,x)} \tilde{\subseteq} F_A$ iff

$x \in \bigcup_{a \in A} G_{(a,x)} \tilde{\subseteq} F_A$ iff $x \in IntF_A$. ∎

Proposition2.1.17[37]

Let $(\tilde{\chi}_A, \tilde{\tau}, A)$ be a soft topological space , and let G_A , F_A be a soft sets over χ , then :

If $F_A \tilde{\subseteq} G_A$ implies that $IntF_A \tilde{\subseteq} IntG_A$.

<u>Theorem 2.1.18[37]</u>

Let $(\tilde{\chi}_A, \tilde{\tau}, A)$ be a soft topological space ,for any two soft sets F_A, G_A over the universe χ, we have :

i) $\quad Int\tilde{\chi}_A = \tilde{\chi}_A \qquad , Int\tilde{\Phi}_A = \tilde{\Phi}_A$

ii) $\quad IntF_A \tilde{\subseteq} F_A$

iii) $\quad Int(F_A \tilde{\cap} G_A) = Int(F_A) \tilde{\cap} Int(G_A)$

iv) $\quad Int(F_A \tilde{\cup} G_A) \tilde{\supseteq} Int(F_A) \tilde{\cup} Int(G_A)$

v) $\quad Int(IntF_A) = IntF_A$

The opposite case of (iv) is incorrect , the following example shows this fact .

<u>Example2.1.19</u>

Let χ be a universal set consisting of five cars , and let $A \subseteq E$ be a subset of a set of parameters which consists of colours , such that :

$$A = \{a_1, a_2, a_3\} \quad , \quad \chi = \{c_1, c_2, c_3, c_4, c_5\} .$$

Note that $F_{1_A} = \{(a_1, \{c_1, c_3\}) , (a_2, \{c_2, c_4\}), (a_3, \{c_3\})\}$

$$F_{2_A} = \{(a_1, \{c_1\}) , (a_2, \{c_2\}), (a_3, \{c_3\})\}$$

$$F_{3_A} = \{(a_2, \{c_3\}) , (a_3, \{c_3\}), (a_1, \{\varphi\})\} , \text{ then we have :}$$

$\tilde{\tau} = \{\tilde{\chi}_A, \tilde{\Phi}_A, F_{1_A}, F_{2_A}, F_{3_A}\}$, form a soft topological space .

Now let $F_A = \{(a_1, \{c_1\}), (a_3, \{c_3\}), (a_2, \varphi)\}$

$$G_A = \{(a_2, \{c_2\}), (a_3, \{c_3\}), (a_1, \varphi)\}$$

$$Int(F_A \tilde{\cup} G_A) = \tilde{\cup} \{K_A \tilde{\in} \tilde{\tau}, \text{ such that } K_A \tilde{\subseteq} F_A \tilde{\cup} G_A\}$$

$$= F_{2_A} \tilde{\cup} F_{3_A}$$

But $IntF_A = \tilde{\cup} \{H_A \tilde{\in} \tilde{\tau} \text{ such that } H_A \tilde{\subseteq} F_A\} = \tilde{\Phi}_A$, and

$IntG_A = \tilde{\Phi}_A$, hence $Int(F_A \tilde{\cup} G_A) \tilde{\nsubseteq} IntF_A \tilde{\cup} IntG_A$.

Definition 2.1.20[38]

Let $(\tilde{\chi}_A, \tilde{\tau}, A)$ be a soft topological space , and let G_A be a soft set over the universe χ , then :

The soft closure of G_A is a soft closed set defined as :

$ClG_A = \tilde{\cap} \{S_A, S_A \text{ is soft closed and } G_A \tilde{\subseteq} S_A\}$.

Theorem 2.1.21

Let $(\tilde{\chi}_A, \tilde{\tau}, A)$ be a soft topological space ,and let F_A be a soft set over the universe χ ,   a soft point $a_F \tilde{\in} ClF_A$ iff for every soft open set G_A of a_F , then :

$$F_A \tilde{\cap} G_A \neq \tilde{\Phi}_A$$

Proof //

Suppose that $a_F \tilde{\in} ClF_A$, and let G_A be a soft open set containing a_F , such that if possible $F_A \tilde{\cap} G_A = \tilde{\Phi}_A$, since G_A is a soft open , then by [Definition 2.1.1] $\tilde{\chi}_A - G_A$ is a soft closed .

Thus $ClF_A \tilde{\subseteq} \tilde{\chi}_A - G_A$ [ClF_A is a smallest soft closed set containing F_A] , but $a_F \tilde{\in} G_A$, a contradiction ,

hence : $F_A \tilde{\cap} G_A \neq \tilde{\Phi}_A$.

conversely :

suppose that the condition holds , and if possible that $a_F \tilde{\notin} ClF_A$.

this means that $a_F \tilde{\in} \tilde{\chi}_A - ClF_A$, and hence $\tilde{\chi}_A - ClF_A$ is a soft open set , thus by the condition $(\tilde{\chi}_A - ClF_A) \tilde{\cap} F_A \neq \tilde{\Phi}_A$, a contradiction.

Hence $a_F \tilde{\in} ClF_A$. ∎

Theorem 2.1.22[38]

Let $(\tilde{\chi}_A, \tilde{\tau}, A)$ be a soft topological space ,and let F_A and G_A be two soft sets over the universe χ, then :

(1) $Cl(\tilde{\Phi}_A) = \tilde{\Phi}_A$, $Cl\tilde{\chi}_A = \tilde{\chi}_A$

(2) $F_A \tilde{\subseteq} G_A$, then $Cl(F_A) \tilde{\subseteq} Cl(G_A)$

(3) $F_A \tilde{\subseteq} Cl(F_A)$

(4) $Cl(F_A \tilde{\cup} G_A) = Cl(F_A) \tilde{\cup} Cl(G_A)$

(5) $Cl(F_A \tilde{\cap} G_A) \tilde{\subseteq} Cl(F_A) \tilde{\cap} Cl(G_A)$

(6) $Cl(Cl(F_A) = Cl(F_A)$

The opposite case of (5) is incorrect , and the following example shows this fact :

Example2.1.23

Let χ be any universal set , such that: $\chi = \{x_1, x_2. x_3\}$ and let $A \subseteq E$ be any subset of a set of parameters such that : $A = \{a_1, a_2\}$.

Now we assume that $F_A = \{(a_1, \{x_1\}), (a_2, \varphi)\}$ and $G_A = \{(a_1, \{x_2\}), (a_2, \{\varphi\})\}$

$$\tilde{\tau} = \{\tilde{\Phi}_A, \tilde{\chi}_A, G_{1_A}, G_{2_A} G_{3_A}\}$$

Where ; $G_{1_A} = \{(a_1, \{x_2\}), (a_2, \{x_3\})$

$$G_{2_A} = \{(a_1, \{x_1\}), (a_2, \{x_1\})$$

$$G_{3_A} = \{(a_1, \{x_1, x_2\}), (a_2, \{x_1, x_3\})$$, then the family of all soft closed set is :

$C(\tilde{\chi}) = \{\tilde{\Phi}_A, \tilde{\chi}_A, \tilde{\chi}_A - G_{1_A}, \tilde{\chi}_A - G_{2_A}, \tilde{\chi}_A - G_{3_A}\}$ such that

$$\tilde{\chi}_A - G_{1_A} = \{(a_1, \{x_1, x_3\}), (a_2, \{x_1, x_2\})\}$$

$$\tilde{\chi}_A - G_{2_A} = \{(a_1, \{x_2, x_3\}), (a_2, \{x_2, x_3\}$$

$$\tilde{\chi}_A - G_{3_A} = \{(a_1, \{x_3\}), (a_2, \{x_2\})\}$$

Now we assume that $F_A = \{(a_1, \{x_1\}), (a_3, \varphi)\}$ and $G_A = \{(a_1, \{x_2\}), (a_3, \varphi)\}$, then :

$Cl(F_A) = \tilde{\chi}_A - G_{1_A}$, and $Cl(G_A) = \tilde{\chi}_A - G_{2_A}$, and $Cl(F_A) \tilde{\cap} Cl(G_A) = (\tilde{\chi}_A - G_{1_A}) \tilde{\cap} (\tilde{\chi}_A - G_{2_A}) = H_A$, where $H_A = \{(a_1, \{x_3\}), (a_2, \{x_2\})\}$, but $F_A \tilde{\cap} G_A = \tilde{\Phi}_A$, and by (1) , $Cl(\tilde{\Phi}_A) = \tilde{\Phi}_A$, hence $Cl(F_A) \tilde{\cap} Cl(G_A) \tilde{\nsubseteq} Cl(F_A \tilde{\cap} G_A)$.

Theorem 2.1.24

Let $(\tilde{\chi}_A, \tilde{\tau}, A)$ be a soft topological space ,and let F_A be a soft set over the universe χ, then for any soft open set G_A then :
$G_A \tilde{\cap} Cl(F_A) \tilde{\subseteq} Cl(G_A \tilde{\cap} F_A)$.

Proof //

Let $a_F \tilde{\in} G_A \tilde{\cap} Cl(F_A)$, then $a_F \tilde{\in} G_A$ and $a_F \tilde{\in} Cl(F_A)$, if possible that $a_F \tilde{\notin} Cl(G_A \tilde{\cap} F_A)$, then by [Theorem 2.1.21] there exists a soft open set H_A of e_F such that $H_A \tilde{\cap} F_A = \tilde{\Phi}_A$, where $(G_A \tilde{\cap} H_A) = K_A$, but $a_F \tilde{\in} Cl(F_A)$, a contradiction with [Theorem 2.1.21] . Hence $a_F \tilde{\in} Cl(G_A \tilde{\cap} F_A)$, that is $G_A \tilde{\cap} Cl(F_A) \tilde{\subseteq} Cl(G_A \tilde{\cap} F_A)$.

Proposition 2.1.25[31]

For any soft set F_A over the universe , then :

(1) $\tilde{\chi}_A - Int(F_A) = Cl(\tilde{\chi}_A - F_A)$
(2) $\tilde{\chi}_A - Cl(F_A) = Int(\tilde{\chi}_A - F_A)$

Corollary 2.1.26

For any soft set F_A over the universe , then :

(1) $Int(F_A) = \tilde{\chi}_A - Cl(\tilde{\chi}_A - F_A)$
(2) $Cl(F_A) = \tilde{\chi}_A - Int(\tilde{\chi}_A - F_A)$

The proof is direct from the definitions and the properties .

Proposition 2.1.27[31]

Let $(\tilde{\chi}_A, \tilde{\tau}, A)$ be a soft topological space , and let F_A , G_A be soft set over χ , then :

1) F_A is soft open iff $Int(F_A) = F_A$
2) G_A is soft closed iff $Cl(G_A) = G_A$

Soft subspace

Let $(\tilde{\chi}_A, \tilde{\tau}, A)$ be a soft topological space , and let $Y \subseteq \chi$, then we may construct a soft s topology $\tilde{\tau}_Y$ for $\tilde{Y}_A$ which is called the relative soft topology or the soft relativization of $\tilde{\tau}$ to $\tilde{Y}_A$, the soft topology defined as follows :

Definition 2.1.28 [38]

Let $(\tilde{\chi}_A, \tilde{\tau}, A)$ be a soft topological space , and let $Y \subseteq \chi$, the $\tilde{\tau}$ −relative soft topology for $\tilde{Y}_A$ is the collection $\tilde{\tau}_Y$, given by :

$$\tilde{\tau}_Y = \{\, G_A \,\tilde{\cap}\, \tilde{Y}_A \,, G_A \in \tilde{\tau}\} \quad , \quad \text{note} \quad \text{that} \quad \tilde{Y}_A \quad \text{means} \quad \text{that}$$
$$\tilde{Y}_A = \{(a, Y\,)\} \qquad \forall a \in A\,.$$

The soft topological space $(\tilde{Y}_A, \tilde{\tau}_Y, A)$ is called soft subspace of $(\tilde{\chi}_A, \tilde{\tau}, A)$, the soft topology $\tilde{\tau}_Y$ is said to be induced by $\tilde{\tau}$.

Example 2.1.29

If we consider [Example2.1.23] . Let χ be any universal set , such that: $\chi = \{x_1, x_2, x_3\}$ and $Y \subseteq \chi = \{x_1, x_2\}$ let $A \subseteq E$ be any subset of a set of parameters such that : $A = \{a_1, a_2\}$ .And

$$\tilde{\tau} = \left\{\tilde{\Phi}_A, \tilde{\chi}_A,\ G_{1_A}\,,\ G_{2_A}\ G_{3_A}\right\}$$

Where ; $\qquad G_{1_A} = \{(a_1, \{x_2\})\,, (a_2, \{x_3\})$

$$G_{2_A} = \{(a_1, \{x_1\})\,, (a_2, \{x_1\})$$

$$G_{3_A} = \{(a_1, \{x_1, x_2\})\,, (a_2, \{x_1, x_3\})\}\ ,\ \text{ then}$$

$$\tilde{\tau}_Y = \{\tilde{Y}_A \,\tilde{\cap}\, G_{1_A}\,, \tilde{Y}_A \,\tilde{\cap}\, G_{2_A}, \tilde{Y}_A \,\tilde{\cap}\, G_{3_A}, \tilde{Y}_A \,\tilde{\cap}\, \tilde{\chi}_A, \tilde{\Phi}_A\,\} \qquad .$$

$$= \left\{\tilde{Y}_A, \tilde{\Phi}_A\,, \{(a_1, \{x_2\})\},\ G_{2_A}\,, \{(a_1, \{x_1, x_2\}), (a_2, \{x_1\})\}\right\}.$$

In order that the above definition may be contestant , we must show that $\tilde{\tau}_Y$ is actually a soft topology for $\tilde{Y}_A$, thus we put the following theorem .

Theorem 2.1.30

Let $(\tilde{\chi}_A, \tilde{\tau}, A)$ be a soft topological space , and let $Y \subseteq \chi$, the collection :

$$\tilde{\tau}_Y = \{\, G_A \,\tilde{\cap}\, \tilde{Y}_A \,,\, G_A \in \tilde{\tau}\,\} \text{ is a soft topology} .$$

The proof is direct . ∎

Theorem 2.1.31

Let $(\tilde{Y}_A, \tilde{\tau}_Y, A)$ be a soft subspace of $(\tilde{\chi}_A, \tilde{\tau}, A)$, and $(\tilde{w}_A, \tilde{\tau}_w, A)$ be a soft subspace of $(\tilde{Y}_A, \tilde{\tau}_Y, A)$, then $(\tilde{w}_A, \tilde{\tau}_w, A)$ be a soft subspace of $(\tilde{\chi}_A, \tilde{\tau}, A)$.

The proof is easy . ∎

Theorem 2.1.32

Let $(\tilde{Y}_A, \tilde{\tau}_Y, A)$ be a soft subspace of $(\tilde{\chi}_A, \tilde{\tau}, A)$, then:

1) Any $\tilde{\tau}_Y$- soft open set G_A is a soft nhd of a soft point e_F iff $G_A = F_A \,\tilde{\cap}\, \tilde{Y}_A$ for some $\tilde{\tau}_Y$- soft open nhd F_A of e_F .
2) For any $G_A \in S_4(Y)$, then $Cl_Y(G_A) = Cl_\chi(G_A) \,\tilde{\cap}\, \tilde{Y}_A$.

Where $Cl_\chi(G_A)\ and\ Cl_Y(G_A)$ is the soft closure with respect to $\tilde{\tau}$ and $\tilde{\tau}_Y$ respectively ,

3) A soft set G_A is $\tilde{\tau}_Y$- soft closed set iff there is a $\tilde{\tau} -$ soft closed F_A such that :

$$G_A = F_A \,\tilde{\cap}\, \tilde{Y}_A$$

Proof (1)//

Suppose that a_F is a soft point and G_A is a $\tilde{\tau}_Y$- soft open *nhd* of a_F then by[definition 2.1.4] there exists a $\tilde{\tau}_Y$- soft open H_A such that $a_F \,\tilde{\in}\, H_A \,\tilde{\subseteq}\, G_A$ iff there is a $\tilde{\tau} -$ soft *open nhd* K_A of a_F such that:

$H_A = K_A \, \widetilde{\cap} \, \widetilde{Y}_A \, \widetilde{\subseteq} \, G_A$, now let $O_A = K_A \, \widetilde{\cup} \, G_A$, then O_A is a $\tilde{\tau} -$ soft *open nhd* of e_F, such that : $a_F \, \widetilde{\in} \, K_A \, \widetilde{\subseteq} \, O_A$, further $O_A \, \widetilde{\cap} \, \widetilde{Y}_A = (K_A \, \widetilde{\cup} \, G_A) \, \widetilde{\cap} \, \widetilde{Y}_A$

$$= (G_A \, \widetilde{\cap} \, \widetilde{Y}_A) \, \widetilde{\cup} \, (K_A \, \widetilde{\cap} \, \widetilde{Y}_A)$$

$$= G_A \, \widetilde{\cup} \, (K_A \, \widetilde{\cap} \, \widetilde{Y}_A) = G_A$$

Conversely:

Let $G_A = O_A \, \widetilde{\cap} \, \widetilde{Y}_A$, for some $\tilde{\tau} -$ soft *open nhd* O_A of a_F , then there is a $\tilde{\tau} -$ soft *open nhd* M_A of a_F *such that* $a_F \, \widetilde{\in} \, M_A \, \widetilde{\subseteq} \, O_A$, which implies that $a_F \, \widetilde{\in} \, M_A \, \widetilde{\cap} \, \widetilde{Y}_A \, \widetilde{\subseteq} \, O_A \, \widetilde{\cap} \, \widetilde{Y}_A = G_A$

Now , since $M_A \, \widetilde{\cap} \, \widetilde{Y}_A$ is a $\tilde{\tau}_Y$- soft open *nhd* of a_F , so G_A is a $\tilde{\tau}_Y$- soft open *nhd* of a_F.

Proof (2)//

Since

$Cl_Y(G_A) = \widetilde{\cap} \, \{K_A \, , K_A$ is soft $\tilde{\tau}_Y$- soft closed set such that $G_A \, \widetilde{\subseteq} \, K_A\}$

$\qquad = \widetilde{\cap} \, \{F_A \, \widetilde{\cap} \, \widetilde{Y}_A, F_A$ is a $\tilde{\tau} -$ soft *closed set* and $G_A \, \widetilde{\subseteq} \, F_A \, \widetilde{\cap} \, \widetilde{Y}_A \}$

$\qquad = \widetilde{\cap} \, \{F_A \, , F_A$ is a $\tilde{\tau} -$soft closed set and $G_A \, \widetilde{\subseteq} \, F_A\} \, \widetilde{\cap} \, \widetilde{Y}_A$

$\qquad = Cl_\chi(G_A) \, \widetilde{\cap} \, \widetilde{Y}_A$

Proof (3)//

Let G_A *be a* $\tilde{\tau}_Y$- soft closed set iff $(\widetilde{Y}_A - G_A)$ be a $\tilde{\tau}_Y$- soft open set iff $(\widetilde{Y}_A - G_A) = F_A \, \widetilde{\cap} \, \widetilde{Y}_A$, for some $\tilde{\tau} -$ soft open F_A [Definition 2.1.28] iff

$G_A = \widetilde{Y}_A \text{-} (F_A \, \widetilde{\cap} \, \widetilde{Y}_A)$ iff $G_A = (\widetilde{Y}_A \text{-} F_A) \, \widetilde{\cup} \, (\widetilde{Y}_A \text{-} \widetilde{Y}_A)$ iff

$G_A = \widetilde{Y}_A \,\widetilde{\cap}\, (\tilde{\chi}_A - F_A)$ iff $G_A = \widetilde{Y}_A \,\widetilde{\cap}\, H_A$, where $H_A = (\tilde{\chi}_A - F_A)$ is a $\tilde{\tau}$ − soft closed set . ∎

Theorem 2.1.33

Let $(\widetilde{Y}_A, \tilde{\tau}_Y, A)$ be a soft subspace of $(\tilde{\chi}_A, \tilde{\tau}, A)$, and F_A *is a soft subset of* $\widetilde{Y}_A$, if F_A is a soft open (closed)set of $\tilde{\tau}$, then it is a soft open (closed)set of $\tilde{\tau}_Y$.

Proof //

$F_A \widetilde{\subseteq} \widetilde{Y}_A$, we have $F_A = F_A \,\widetilde{\cap}\, \widetilde{Y}_A$ so that F_A is a soft intersection of $\widetilde{Y}_A$ with a soft open (closed)set of $\tilde{\tau}$, hence by [Definition 2.1.28]] , F_A is a soft open (closed)set of $\tilde{\tau}_Y$. ∎

Definition 2.1.34 [31]

A soft mapping f_{pu} from a soft topological space $(\tilde{\chi}_A, \tilde{\tau}, B)$ into a soft topological space$(\tilde{Y}_B, \tilde{\rho}, B)$ is a soft continuous iff $f_{pu}^{-1}(G_A) \,\tilde{\in}\, \tilde{\tau}$ for any $G_A \,\tilde{\in}\, \tilde{\rho}$.

Definition 2.1.35

Let $(\tilde{\chi}_A, \tilde{\tau}, A)$ and $(\tilde{Y}_B, \tilde{\rho}, B)$ be two soft topological spaces overχ and Y respectively , then for $a \in A$ and $x \in \chi$, the map $f_{pu}: (\tilde{\chi}_A, \tilde{\tau}, A) \longrightarrow (\tilde{Y}_B, \tilde{\rho}, B)$ is an a −soft continuous at x iff for each $p(a)$ −soft open set G_B containing $u(x)$ there exists an a −soft open set H_A containing x, such that $f_{pu}(H_A) \,\tilde{\subseteq}_{p(a)}\, G_B$.

Definition 2.1.36[39]

Let $(\tilde{\chi}_A, \tilde{\tau}, A)$ and $(\tilde{Y}_B, \tilde{\rho}, B)$ be two soft topological spaces over χ and Y respectively , the map $f_{pu}: (\tilde{\chi}_A, \tilde{\tau}, A) \longrightarrow (\tilde{Y}_B, \tilde{\rho}, B)$ is called soft open map if for each soft open set G_A in $\tilde{\chi}_A$, then $f_{pu}(G_A)$ is soft open set in $\tilde{Y}_B$.

Definition 2.1.37

Let $(\tilde{\chi}_A, \tilde{\tau}, A)$ and $(\tilde{Y}_B, \tilde{\rho}, B)$ be two soft topological spaces over χ and Y respectively , the map $f_{pu}: (\tilde{\chi}_A, \tilde{\tau}, A) \longrightarrow (\tilde{Y}_B, \tilde{\rho}, B)$ is called soft closed map if for each soft closed set G_A in $\tilde{\chi}_A$ then $f_{pu}(G_A)$ is soft closed set in $\tilde{Y}_B$.

Definition 2.1.38

A map $f_{pu}: (\tilde{\chi}_A, \tilde{\tau}, A) \longrightarrow (\tilde{Y}_B, \tilde{\rho}, B)$ is called soft bi continuous iff f_{pu} is soft open and soft continuous .

Definition 2.1.39

A map $f_{pu}: (\tilde{\chi}_A, \tilde{\tau}, A) \longrightarrow (\tilde{Y}_B, \tilde{\rho}, B)$ is called soft homeomorphism iff :

a) f_{pu} is soft bijective , that is [both of $p \; and \; u$ are bijective].

b) f_{pu} is soft continuous .

c) f_{pu}^{-1} is soft continuous .

(2.2) Soft (limits , boundary , exterior) points .

In this section , we will study the soft limit points (soft accumulation points) in addition to the soft boundary points , soft exterior points , the properties of them and some results .

Definition 2.2.1

Let $(\tilde{X}_A, \tilde{\tau}, A)$ be a soft topological space , a soft point a_F of a soft set F_A is said to be a soft limit point of F_A iff for every soft nhd G_A of a_F ,then :

$$F_A \,\tilde{\cap}\, G_A \backslash a_F \neq \tilde{\Phi}_A$$

- The set of all soft limit points of a soft set F_A is called soft derived set of F_A , and shall be denoted by $D(F_A)$.

It is easy , a_F is not soft limit point of F_A if there is a soft *nhd* G_A of a_F such that

$$F_A \,\tilde{\cap}\, G_A \backslash a_F = \tilde{\Phi}_A$$

Example2.2.2

Assume that the universal set χ consisting of four types of different animals such that $\chi = \{p_1, p_2, p_3, p_4\}$, where :

$p_1 = Loin$, $p_2 = Rabbite$, $p_3 = Fox$, $p_4 = wolf$.

And suppose that the subset of parameters A of E consisting of :

$$A = \{a_1 = plants , a_2 = meat \}$$

The soft set F_A over χ is defined as :

$F_A = \{(a_1, \{p_2\}), (a_2, \{p_1, p_2, p_3\})\}$, and :

$\tilde{\tau} = \{\tilde{\Phi}_A, \tilde{X}_A, G_{1_A}, G_{2_A}\}$, where $G_{1_A} = \{(a_1, \{p_2\}), (a_2, \{p_3\})\}$

$$G_{2_A} = \{(a_1, \{p_2\}), (a_2, \{p_3, p_4\})\}$$

Now $a_F = \{(a_1, \{p_2\}), (a_2, \varphi)\}$, be a soft point of F_A and since for all $G_{i_A} \,\tilde{\in}\, \tilde{\tau}$, $i = 1,2$ then :

$F_A \, \widetilde{\cap} \, G_{i_A} \backslash a_F \neq \widetilde{\Phi}_A$, hence a_F is a soft limit point of F_A .

Definition2.2.3

Let $(\widetilde{\chi}_A, \widetilde{\tau}, A)$ be a soft topological space , for $a \in A \ and \ x \in \chi$ an a −soft point x_a is said to be a −soft limit point (a −soft drived point) of a soft set F_A iff for every a −soft nhd $G_{(a,x)}$ of

x_a , then :

$$F_A \, \widetilde{\cap}_a \, G_{(a,x)} \backslash x_a \neq \widetilde{\Phi}_a$$

• The set of all a −soft limit point of F_A is denoted by $D(F, a)$.

x_a is not a −soft limit point of F_A if there exists an a −soft nhd $G_{(a,x)}$ of x_a such that : $F_A \, \widetilde{\cap}_a \, G_{(a,x)} \backslash x_a = \widetilde{\Phi}_a$.

Example2.2.4

If we consider [Example 2.2.2] then for the point $a_1 \in A$, $p_2 \in \chi$, then , $F_A = \{(a_1, \{p_2\}), (a_2, \{p_1, p_2, p_3\})\}$, and :

$$\widetilde{\tau} = \{\widetilde{\Phi}_A, \widetilde{\chi}_A, G_{1_A}, G_{2_A}\}, \text{ where } G_{1_A} = \{(a_1, \{p_2\}), (a_2, \{p_3\})\}$$

$$G_{2_A} = \{(a_1, \{p_2\}), (a_2, \{p_3, p_4\})\}$$

$p_{2_{a_1}} \in_{a_1} F_A$ and for each a_1 −soft open set $G_{(a_1,p_2)}$ containing p_2 , then

$F_A \, \widetilde{\cap}_a \, G_{(a_1,p_2)} \backslash p_{2_{a_1}} \neq \widetilde{\Phi}_{a_1}$, hence $p_{2_{a_1}}$ is an a_1 −soft limit point .

Definition2.2.5[44]

Let $(\widetilde{\chi}_A, \widetilde{\tau}, A)$ be a soft topological space , a soft point a_F is said to be soft cluster point of a soft set F_A if for every soft open *nhd* G_A of a_F , we have :
$F_A \, \widetilde{\cap} \, G_A \neq \widetilde{\Phi}_A$.

The set of all soft cluster points of F_A is denoted by $Sdh(F_A)$.

Definition 2.2.6

Let $(\tilde{\chi}_A, \tilde{\tau}, A)$ be a soft topological space , a soft point a_F of a soft set F_A is said to be soft isolated point of F_A iff there is a soft open set G_A of a_F such that :

$$F_A \,\tilde{\cap}\, G_A \backslash a_F = \tilde{\Phi}_A$$

Example 2.2.7

Assume that $A \subseteq E$ a subset of parameters such that $A = \{a_1 , a_2\}$ and let χ be a universal set such that $\chi = \{x_1, x_2, x_3\}$,

if $F_A = \{(a_1, \{x_1\}), (a_2, \{x_1, x_2\})\}$ be a soft set over χ .

let $\quad a_F = \{ \quad (a_1, \{x_1\}), (a_2, \varphi) \} \quad$ be a soft point and $\tilde{\tau} = \{\tilde{\Phi}_A, \tilde{\chi}_A, F_{1_A}, F_{2_A}, F_{3_A} \}$ be a soft topological space , where :

$$F_{1_A} = \{(a_1, \{x_1, x_2\}), (a_2, \{x_1, x_3\})\}$$

$$F_{2_A} = \{(a_1, \{x_3\}),(a_2, \{x_1, x_3\})\}$$

$$F_{3_A} = \{(a_2, \{x_1\})\} \text{ , not that } a_F \text{ is a soft isolated point}$$
of F_A , because there is a soft open set F_{3_A} such that :

$$F_A \,\tilde{\cap}\, F_{3_A} \backslash a_F = \tilde{\Phi}_A.$$

Definition 2.2.8

Let $(\tilde{\chi}_A, \tilde{\tau}, A)$ be a soft topological space , for $a \in A$ and $x \in \chi$, F_A be a soft set over χ , an a −soft point x_a is said to be an a −soft solated point of a soft set F_A iff there exists a soft open set $G_{(a,x)}$ of x_a such that:

$$F_A \,\tilde{\cap}_a\, G_{(a,x)} \backslash x_a = \tilde{\Phi}_a.$$

Example2.2.9

If we consider example [2.2.7] then for a point $a_2 \in A$ and $x_1 \in \chi$

$F_A = \{(a_1, \{x_1\}), (a_2, \{x_1, x_2\})\}$, note $x_{1_{a_2}} = \{(a_2, \{x_1\})\}$ is an a_2

-soft point of F_A .

Now since F_{3_A} is an a_2 -soft point of F_A such that :

$F_A \,\widetilde{\cap}_a\, F_{3_A} \setminus x_{1_{a_2}} = \widetilde{\Phi}_{a.}$, hence $x_{1_{a_2}}$ is a_2-soft isolated point of F_A .

<u>Definition2.2.10</u>

Let $(\tilde{\chi}_A, \tilde{\tau}, A)$ be a soft topological space, a soft closed set F_A which has no soft isolated point is called perfect soft set .

<u>Example 2.2.11</u>

Suppose that we have the universe set χ consisting of five coins as follows :

$$\chi = \{x_1 = (doular)\,, x_2 = (Euro), x_3 = (pound), x_4 = (dinar), x_5 = (dirham)\}$$

And we have a set of parameters $A \subseteq E$ consisting of three types of cars , as :

$$A = \{a_1 = (BMW), a_2 = (mercedes), a_3 = (Toyota)\}$$

The mapping between A and $IP(\chi)$ indicates the ability of Mr.X_i $i = 1,2,3,4,5$ to buy these cars :

Consider the following soft topology :

$$\tilde{\tau} = \{\widetilde{\Phi}_A, \tilde{\chi}_A, G_{1_A}, G_{2_A}, G_{3_A}, G_{4_A}, G_{5_A}\}$$

Such that : $G_{1_A} = \{(a_1, \{x_1, x_2, x_3\}), (a_2, \{x_3, x_4\}), (a_3, \varphi)\}$

$\qquad\qquad G_{2_A} = \{(a_1, \{x_1, x_2\}), (a_3, \{x_1, x_2, x_5\}), (a_3, \varphi)\}$

$\qquad\qquad G_{3_A} = \{(a_3, \{x_2, x_5\}), (a_1, \varphi), (a_2, \varphi)\}$

$\qquad\qquad G_{4_A} = \{(a_1, \{x_1\}), (a_2, \varphi), (a_3, \varphi)\}$

$\qquad\qquad G_{5_A} = \{(a_2, \{x_2\}), (a_1, \varphi), (a_3, \varphi)\}$

Now since $\tilde{\chi}_A$ is a soft closed set and $\forall\, a_F \,\widetilde{\in}\, \tilde{\chi}_A,$ and $\forall\, G_{i_A} \,\widetilde{\in}\, \tilde{\tau}$ $i = 1,2,3,4,5$, then :

$\tilde{\chi}_A \backslash a_F \,\tilde{\cap}\, G_{i_A} \neq \tilde{\Phi}_A$, then $\tilde{\chi}_A$ has no soft isolated points , and hence it a perfect soft set .

Definition2.2.12

Let $(\tilde{\chi}_A, \tilde{\tau}, A)$ be a soft topological space , for $a \in A$, an $a -$ soft closed set F_A which has no $a -$ soft isolated point is called $a -$ perfect soft set .

Example2.2.13

If we consider example [2.2.11] then :

For $a \in A$ and $x \in \chi$, $\forall\, x_a \,\tilde{\subseteq}_a\, \tilde{\chi}_a$ and $\forall\, G_{i_A} \,\tilde{\in}\, \tilde{\tau}$ of x_a, we have ;

$$\tilde{\chi}_a \backslash x_a \,\tilde{\cap}_a\, G_{i_A} \neq \tilde{\Phi}_a$$

Thus $\tilde{\chi}_a$ has no $a -$soft isolated points , hence $\tilde{\chi}_a$ is an $a -$perfect soft set .

Theorem 2.2.14

Let $(\tilde{\chi}_A, \tilde{\tau}, A)$ be a soft topological space , a soft set F_A is soft closed set iff $D(F_A) \,\tilde{\subseteq}\, F_A$.

Proof//

Let F_A be a soft closed set , then by [Definition 2.1.1] $(\tilde{\chi}_A - F_A)$ is soft open set and so ; for each soft point $a_F \,\tilde{\in}\, (\tilde{\chi}_A - F_A), a_F \,\tilde{\notin}\, F_A$, there is a soft open nhd G_A of a_F such that :

$$G_A \,\tilde{\subseteq}\, (\tilde{\chi}_A - F_A)$$

And since $F_A \,\tilde{\cap}\, (\tilde{\chi}_A - F_A) = \tilde{\Phi}_A$ [definition 2.2.1] , a_F is not soft limit point of F_A , thus $G_A \,\tilde{\cap}\, F_A = \tilde{\Phi}_A$, which implies that $a_F \,\tilde{\notin}\, D(F_A)$ [definition 2.2.1].

Hence F_A contains all its soft limit point thus $D(F_A) \,\tilde{\subseteq}\, F_A$.

Conversely :

assume that $a_F \,\tilde{\notin}\, D(F_A)$, implis that a_F is not soft limit point , hence by [Definition 2.2.1] there is a soft open nhd G_A such that ;

$G_A \widetilde{\cap} F_A = \widetilde{\Phi}_A$ which means that $G_A \widetilde{\subseteq} (\widetilde{\chi}_A - F_A)$ and by .
$(\widetilde{\chi}_A - F_A)$ is soft open set , there for F_A is soft closed set . ∎

Theorem 2.2.15

Let $(\widetilde{\chi}_A, \widetilde{\tau}, A)$ be a soft topological space , and let F_A , G_A be a soft
sets over the universe χ , then the following properties hold :

(i) $D(\widetilde{\Phi}_A) = \widetilde{\Phi}_A$
(ii)   If $F_A \widetilde{\subseteq} G_A$, then $D(F_A) \widetilde{\subseteq} D(G_A)$
(iii) $D(F_A \widetilde{\cap} G_A) \widetilde{\subseteq} D(F_A \widetilde{\cap} (G_A)$
(iv) $D(F_A \widetilde{\cup} G_A) = D(F_A) \widetilde{\cup} (G_A)$

Proof //(i) obviously

Proof //(ii) assume that $D(F_A)$ be a soft set containing a soft point a_F ,
a_F be a soft limit point , then by [Definition 2.2.1] , for each soft nhd
R_A of a_F , then :
$F_A \widetilde{\cap} R_A \backslash a_F \neq \widetilde{\Phi}_A$, but $F_A \widetilde{\subseteq} G_A$, hence $G_A \widetilde{\cap} R_A \backslash a_F \neq \widetilde{\Phi}_A$, this means
that : $a_F \widetilde{\in} D(G_A)$, so $D(F_A) \widetilde{\subseteq} D(G_A)$

Proof //(iii) since $F_A \widetilde{\cap} G_A \widetilde{\subseteq} F_A$, then by [2] $D(F_A \widetilde{\cap} G_A) \widetilde{\subseteq} D(F_A)$
...............(2.3)
$F_A \widetilde{\cap} G_A \widetilde{\subseteq} G_A$, implies $D(F_A \widetilde{\cap} G_A) \widetilde{\subseteq} D(G_A)$ ....(2.4)
From (1) and (2) $D(F_A \widetilde{\cap} G_A) \widetilde{\subseteq} D(F_A \widetilde{\cap} (G_A)$.

Proof //(iv) let $a_F \widetilde{\in} D(F_A) \widetilde{\cup} (G_A)$, then

either $a_F \widetilde{\in} D(F_A)$ or $a_F \widetilde{\in} D(G_A)$, then there is a soft open *nhd* K_A of
a_F such that : $F_A \widetilde{\cap} K_A \backslash a_F \neq \widetilde{\Phi}_A$ or $G_A \widetilde{\cap} K_A \backslash a_F \neq \widetilde{\Phi}_A$, this implies that
$(F_A \widetilde{\cup} G_A) \widetilde{\cap} K_A \backslash a_F \neq \widetilde{\Phi}_A$, i.e $a_F \widetilde{\in} D(F_A \widetilde{\cup} G_A)$, hence

$$D(F_A \widetilde{\cup} G_A) \widetilde{\subseteq} D(F_A) \widetilde{\cup} (G_A)$$

Conversely

Since $F_A \widetilde{\subseteq} F_A \widetilde{\cup} G_A$, $G_A \widetilde{\subseteq} F_A \widetilde{\cup} G_A$, then by (ii) $D(F_A) \widetilde{\subseteq} D(F_A \widetilde{\cup} G_A)$ and
$D(G_A) \widetilde{\subseteq} D(F_A \widetilde{\cup} G_A)$, thus $D(F_A) \widetilde{\cup} (G_A) \widetilde{\subseteq} D(F_A \widetilde{\cup} G_A)$.

so $D(F_A \widetilde{\cup} G_A) = D(F_A) \widetilde{\cup} (G_A)$. ∎

Example 2.2.16

if we consider the soft topology in example [2.2.7] , and

$$F_A = \{(a_1, \{x_1, x_2\}), (a_2, \{\chi\})\}$$

$G_A = \{(a_1, \{x_1\}), (a_2, \{\chi\})\}$, note that $F_A \cap G_A = \{(a_2, \{\chi\})\}$, and $\{(a_2, \{\chi\})\} \tilde{\notin} D(F_A \cap G_A)$, thus $D(F_A \cap G_A) = \tilde{\Phi}_A$................(2.5)

Now since $a_{1_F} = \{(a_1, \{x_1, x_2\})\}$, and $a_{2_F} = \{(a_2, \{x_1\})\}$ are soft point of F_A , and F_{2_A} is the soft *nhd* of a_{1_F} such that :

$F_A \cap F_{2_A} \backslash a_{1_F} \neq \tilde{\Phi}_A$, thus $a_{1_F} \tilde{\in} D(G_A)$

Also $a_{2_F} = \{(a_2, \{x_1\})\}$ such that G_{2_A} is the soft *nhd* containing a_{2_F} , such that : $F_A \cap G_{2_A} \backslash a_{2_F} \neq \tilde{\Phi}_A$, thus $a_{2_F} \tilde{\in} D(F_A)$,

Hence $D(F_A) = FA$

Now , $a_{2_G} = \{(a_2, \{\chi\})\}$ and $a_{2_G} = \{(a_2, \{x_1\})\}$, are the soft point of G_A , and there exists $G_{2_A} \tilde{\in} \tilde{\tau}$ such that $G_A \cap F_{2_A} \backslash a_{2_G} = \tilde{\Phi}_A$

Thus $a_{2_F} \tilde{\notin} D(G_A)$, and since $\tilde{\chi}_A$ is the unique soft *nhd* containing a_{2_F} such that : $\tilde{\chi}_A \backslash \{a_{2_F}\} \cap G_A \neq \tilde{\Phi}_A$, thus

$a_{2_G} \tilde{\in} D(G_A)$ hence $D(F_A) \cap D(G_A) = \{(a_2, \{x_1\}), (a_2, \varphi)\} \neq \tilde{\Phi}_A$

Thus we have $D(F_A) \cap D(G_A) \tilde{\nsubseteq} D(F_A \cap G_A)$.

Theorem 2.2.17

For any soft set F_A over the universe χ , then :

$$ClF_A = F_A \tilde{\cup} D(F_A)$$

Proof//

We are first prove that $F_A \tilde{\cup} D(F_A)$ is a closed soft set , that is $\tilde{\chi}_A - (F_A \tilde{\cup} D(F_A)) = (\tilde{\chi}_A - F_A) \cap (\tilde{\chi}_A - D(F_A))$ is soft open set .

Now for a soft point $a_F \tilde{\in} (\tilde{\chi}_A - F_A) \cap (\tilde{\chi}_A - D(F_A))$, then:

$a_F \tilde{\in} (\tilde{\chi}_A - F_A)$ *and* $a_F \tilde{\in} \tilde{\chi}_A - D(F_A)$, thus $a_F \tilde{\notin} F_A$ and $a_F \tilde{\notin} D(F_A)$

So by [Definition 2.2.1] , there is a soft nhd R_A of a_F such that :

$R_A \,\tilde{\cap}\, F_A = \widetilde{\Phi}_A$, hence $R_A \,\widetilde{\subseteq}\, \tilde{\chi}_A - F_A$, now for each $a_G \,\widetilde{\in}\, R_A$, then $a_G \,\widetilde{\notin}\, D(F_A)$, then $R_A \,\tilde{\cap}\, D(F_A) = \widetilde{\Phi}_A$, this implies that $R_A \,\widetilde{\subseteq}\, \tilde{\chi}_A - D(F_A)$ (i.e $R_A \,\widetilde{\subseteq}\, (\tilde{\chi}_A - F_A) \,\tilde{\cap}\, (\tilde{\chi}_A - D(F_A))$)

Thus $(\tilde{\chi}_A - F_A) \,\tilde{\cap}\, (\tilde{\chi}_A - D(F_A))$ is a soft nhd of all its elements , and hence $(\tilde{\chi}_A - F_A) \,\tilde{\cap}\, (\tilde{\chi}_A - D(F_A))$ is a soft open set and thus $F_A \,\tilde{\cup}\, D(F_A)$ is a soft closed set containing F_A , there fore $ClF_A \,\widetilde{\subseteq}\, F_A \,\tilde{\cup}\, D(F_A)$,

And since ClF_A is a soft closed [Definition 2.1.20] and then by [Theorem 2.2.14] ClF_A contains of all its soft limits points .

Also it is easy ClF_A contains all soft limits points F_A $[F_A \,\widetilde{\subseteq}\, ClF_A]$

Thus $D(F_A) \,\widetilde{\subseteq}\, ClF_A$ and $F_A \,\widetilde{\subseteq}\, ClF_A$, hence
$$ClF_A = F_A \,\tilde{\cup}\, D(F_A). \ \blacksquare$$

Theorem 2.2.18

Let $(\tilde{\chi}_A, \tilde{\tau}, A)$ be a soft topological space , and let F_A be a soft set over the universe χ , then $IntF_A = \{a_F \,\widetilde{\in}\, F_A \ ; a_F \,\widetilde{\notin}\, D(\tilde{\chi}_A - F_A)\}$.

Proof//

Let a_F be a soft point of F_A which is not soft limit points of $\tilde{\chi}_A - F_A$, then by [Definition 2.2.1] there is a soft nhd G_A of a_F which does note contain a point of $\tilde{\chi}_A - F_A$.

And so $G_A \,\widetilde{\subseteq}\, F_A$, this implies that F_A is also soft *nhd* of a_F and so by [Definition 2.1.12] , a_F is soft interior point of F_A .

Conversely :

Let $a_F \,\widetilde{\in}\, Int(F_A)$ then $\exists G_A \,\widetilde{\in}\, \tilde{\tau}$ such that $a_F \,\widetilde{\in}\, G_A \,\widetilde{\subseteq}\, F_A$, this implies that $G_A \,\tilde{\cap}\, \tilde{\chi}_A - F_A = \widetilde{\Phi}_A$, that is $a_F \,\widetilde{\notin}\, \tilde{\chi}_A - F_A$ $\quad \blacksquare$

Definition 2.2.19

Let $(\tilde{\chi}_A, \tilde{\tau}, A)$ be a soft topological space ,and let F_A be a soft set over the universe χ , , a soft point a_F is said to be soft boundary point of a soft set F_A iff for each soft nhd G_A of a_F , then :

$$F_A \, \widetilde{\cap} \, G_A \neq \widetilde{\Phi}_A \text{ , and } (\widetilde{\chi}_A - F_A) \, \widetilde{\cap} \, G_A \neq \widetilde{\Phi}_A$$

The set of all soft boundary points is denoted by $Sb(F_A)$.

Example2.2.20

Assume that $A \subseteq E$ a subset of parameters such that $A = \{a_1 , a_2\}$ and let χ be a universal set such that $\chi = \{x_1, x_2, x_3\}$, let :

$$F_A = \{(a_1, \{x_1\}), (a_2, \{x_3\})\} \text{ and}$$

$$\widetilde{\tau} = \{\widetilde{\Phi}_A, \widetilde{\chi}_A, \, G_{1_A} , \, G_{2_A} \, G_{3_A}\}$$

Where ;
$$G_{1_A} = \{(a_1, \{x_2\}) , (a_2, \{x_3\})\}$$

$$G_{2_A} = \{(a_1, \{x_1\}) , (a_2, \{x_1\})\}$$

$$G_{3_A} = \{(a_1, \{x_1, x_2\}) , (a_2, \{x_1, x_3\})$$

For $a_F = \{(a_1, \{x_1\})\} \, \widetilde{\in} \, F_A$ then for any soft open set G_A containing a_F , then : $F_A \, \widetilde{\cap} \, G_A \neq \widetilde{\Phi}_A$, and $(\widetilde{\chi}_A - F_A) \, \widetilde{\cap} \, G_A \neq \widetilde{\Phi}_A$. Thus a_F is soft boundary point of F_A .

Note 2.2.21

1. For any soft set F_A , a soft point a_F is a soft boundary point of F_A iff it is not a soft interior point of F_A nor soft interior point of $(\widetilde{\chi}_A - F_A)$.
2. $Sb(F_A) = Sb(\widetilde{\chi}_A - F_A)$,is direct from the definition .

Theorem 2.2.22

Let $(\widetilde{\chi}_A, \widetilde{\tau}, A)$ be a soft topological space, For any soft set F_A over the universe χ , we have :

$$Sb(F_A) = Cl(F_A) \, \widetilde{\cap} \, \left(Cl(\widetilde{\chi}_A - F_A) \right)$$

Proof//

$a_F \, \widetilde{\in} \, Sb(F_A)$ iff $F_A \, \widetilde{\cap} \, G_A \neq \widetilde{\Phi}_A$, and $(\widetilde{\chi}_A - F_A) \, \widetilde{\cap} \, G_A \neq \widetilde{\Phi}_A$ for each soft open set G_A [Definition 2.2.19] iff $a_F \, \widetilde{\in} \, Cl(F_A)$ and $a_F \, \widetilde{\in} \, Cl(\widetilde{\chi}_A - F_A)$ [Theorem 2.1.21] . ∎

Theorem 2.2.23

Let $(\tilde{\chi}_A, \tilde{\tau}, A)$ be a soft topological space ,and let F_A be a soft set over the universe χ , then the following statements are true :

(1) $ClF_A = F_A \,\tilde{\cup}\, Sb(F_A)$

(2) $ClF_A = Int(F_A) \,\tilde{\cup}\, Sb(F_A)$

(3) F_A is a soft closed set iff $Sb(F_A) \,\tilde{\subseteq}\, F_A$

(4) F_A is soft open iff $Sb(F_A) \,\tilde{\subseteq}\, \tilde{\chi}_A - F_A$

Proof //(1)

$Sb(F_A) = Cl(F_A) \,\tilde{\cap}\, Cl(\tilde{\chi}_A - F_A)$

$F_A \,\tilde{\cup}\, Sb(F_A) = F_A \,\tilde{\cup}\, [Cl(F_A) \,\tilde{\cap}\, Cl(\tilde{\chi}_A - F_A) \,]$

$= F_A \,\tilde{\cup}\, Cl(F_A) \,\tilde{\cap}\, [F_A \,\tilde{\cup}\, Cl(\tilde{\chi}_A - F_A)]$

$= Cl(F_A) \,\tilde{\cap}\, \tilde{\chi}_A = Cl(F_A)$.

Proof //(2)

$\qquad Int(F_A) \,\tilde{\cup}\, Sb(F_A) = Int(F_A) \,\tilde{\cup}\, (Cl(F_A) \,\tilde{\cap}\, Cl(\tilde{\chi}_A - F_A))$

$= [Int(F_A) \,\tilde{\cup}\, Cl(F_A)] \,\tilde{\cap}\, [Int(F_A) \,\tilde{\cup}\, Cl(\tilde{\chi}_A - F_A)]$

$= Cl(F_A) \,\tilde{\cap}\, \tilde{\chi}_A = Cl(F_A)$

Proof //(3)

Let F_A be a soft closed set iff $F_A = Cl(F_A)$, and by (2) $Sb(F_A) \,\tilde{\subseteq}\, Cl(F_A)$, hence $Sb(F_A) \,\tilde{\subseteq}\, F_A$.

Proof //(4)

Since F_A be a soft open set , then by [Definition 2.1.1_] $\tilde{\chi}_A - F_A$ be a soft closed set iff [by 5] $Sb(\tilde{\chi}_A - F_A) \,\tilde{\subseteq}\, \tilde{\chi}_A - F_A$, but $Sb(\tilde{\chi}_A - F_A) = Sb(F_A)$, [evidently from 1] , thus $Sb(F_A) \,\tilde{\subseteq}\, \tilde{\chi}_A - F_A$. ∎

Definition 2.2.24

Let $(\tilde{\chi}_A, \tilde{\tau}, A)$ be a soft topological space ,and let F_A be a soft set over the universe χ , a soft point a_F is said to be soft exterior point of F_A iff it is a soft interior point of $\tilde{\chi}_A - F_A$.

That is there is a soft open set G_A of a_F such that $a_F \,\tilde{\in}\, G_A \,\tilde{\subseteq}\, \tilde{\chi}_A - F_A$.

Or equivalently $a_F \,\tilde{\in}\, G_A$ and , $\qquad\qquad F_A \,\tilde{\cap}\, G_A = \tilde{\Phi}_A.$

We will denoted by $ext(F_A)$, as a soft exterior points of F_A , from the Definition [2.1.12] we have :

1. $extF_A = Int(\tilde{\chi}_A - F_A)$
2. $F_A \, \tilde{\cap} \, extF_A = \widetilde{\Phi}_A$, That is no soft point of F_A can be a soft exterior points of F_A .
3. $extF_A = Int(\tilde{\chi}_A - F_A) = \tilde{\chi}_A - Cl(F_A)$, that is a_F is soft exterior point of F_A iff it is not soft cluster point of F_A .

Theorem 2.2.25

Let $(\tilde{\chi}_A, \tilde{\tau}, A)$ be a soft topological space ,for any two soft sets F_A and G_A ,then:

    (1) $ext(\tilde{\chi}_A) = \widetilde{\Phi}_A$, $ext(\widetilde{\Phi}_A) = \tilde{\chi}_A$

    (2) $ext(F_A) = Int(\tilde{\chi}_A - F_A)$

    (3) $ext(F_A) = ext(\tilde{\chi}_A - extF_A)$

    (4)   If $F_A \tilde{\subseteq} G_A$, then $ext(G_A) \tilde{\subseteq} ext(F_A)$

    (5) $Int(F_A) \tilde{\subseteq} ext(extF_A)$

    (6) $ext(F_A \, \tilde{\cup} \, G_A) = ext(F_A) \, \tilde{\cap} \, ext(G_A)$

Proof//(1) and (2) are directly .

Proof//(3)

$$ext(\tilde{\chi}_A - ext(F_A)) = ext(\tilde{\chi}_A - Int(\tilde{\chi}_A - F_A)) \quad =$$

$$= Int\left(\tilde{\chi}_A - [(\tilde{\chi}_A - Int(\tilde{\chi}_A - F_A)))\right)$$

$$= Int(Int(\tilde{\chi}_A - F_A))$$

$$= Int(\tilde{\chi}_A - F_A)$$

$$= ext(F_A)$$

Proof//(4) directly .

Proof//(5)

By (2) we have $ext(F_A) \tilde{\subseteq} \tilde{\chi}_A - F_A$, then (4) gives

$ext(\tilde{\chi}_A - F_A) \tilde{\subseteq} ext(ext(F_A))$, but $Int(F_A) = ext(\tilde{\chi}_A - F_A)$,

Hence $Int(F_A) \tilde{\subseteq} ext(ext(F_A))$.

Proof//(6)

$$ext(F_A \,\tilde{\cup}\, G_A) = Int(\tilde{\chi}_A - (F_A \,\tilde{\cup}\, G_A)) = Int((\tilde{\chi}_A - F_A) \,\tilde{\cap}\, (\tilde{\chi}_A - G_A))$$

$$= ext(F_A) \,\tilde{\cap}\, ext(G_A) . \blacksquare$$

Theorem 2.2.26

Let $(\tilde{\chi}_A, \tilde{\tau}, A)$ be a soft topological space , and let F_A be a soft set over the universe χ, then $IntF_A$, $extF_A$ and $Sb(F_A)$,are soft disjoint ,

i.e $IntF_A \,\tilde{\cap}\, extF_A \,\tilde{\cap}\, Sb(F_A) = \tilde{\Phi}_A$, and $\tilde{\chi}_A = IntF_A \,\tilde{\cup}\, extF_A \,\tilde{\cup}\, Sb(F_A)$, further $Sb(F_A)$ is a soft closed set .

proof//

by definition of $extF_A = Int(\tilde{\chi}_A - F_A)$, also $IntF_A \,\tilde{\subseteq}\, F_A$ and $Int(\tilde{\chi}_A - F_A) \tilde{\subseteq} \tilde{\chi}_A - F_A$, now , since $F_A \,\tilde{\cap}\, (\tilde{\chi}_A - F_A) = \tilde{\Phi}_A$, it follows that $IntF_A \,\tilde{\cap}\, extF_A = \tilde{\Phi}_A$, $IntF_A \,\tilde{\cap}\, extF_A \,\tilde{\cap}\, Sb(F_A) = \tilde{\Phi}_A \,\tilde{\cap}\, Sb(F_A) = \tilde{\Phi}_A$

now , we have $a_F \,\tilde{\in}\, Sb(F_A)$ iff$a_F \,\tilde{\notin}\, Int(F_A)$ and $a_F \,\tilde{\notin}\, extF_A$ iff $a_F \,\tilde{\in}\, [\tilde{\chi}_A - [IntF_A \,\tilde{\cup}\, extF_A]$,thus $Sb(F_A) = [\tilde{\chi}_A - [IntF_A \,\tilde{\cup}\, extF_A]$. (2.6) . $\tilde{\chi}_A = Sb(F_A) \,\tilde{\cup}\, (\tilde{\chi}_A - Sb(F_A)) = Sb(F_A) \,\tilde{\cup}\, Int(F_A) \,\tilde{\cup}\, Sb(F_A)$

But $IntF_A$ and $extF_A$ are soft open sets , we see that from (2.6) that $Sb(F_A)$ is a soft closed set . $\blacksquare$

Note 2.2.27

by the same method of the proof of [Proposition 2.1.16] we can proof that all the following cases :

(1) $\bigcup_{a\in A} a - ClF_A = ClF_A$
(2) $\bigcup_{a\in A} a - SbF_A = SbF_A$
(3) $\bigcup_{a\in A} a - extF_A = extF_A.$

Where the concepts of $a - ClF_A$, $-extF_A$ and $a - SbF_A$ are defined as $a - Int(F_A)$.

Chapter three

Soft Resolvable Space

Introduction

This chapter regards the aim of the thesis , it contains the important concepts that we worked to introduce , soft density and soft resolvability in addition to many new definitions and properties which , in turn , lead to appearing a new theorems as applications to it .

(3.1)Density in soft topological spaces

In this section we will introducing many of a new definitions , properties and results which depends on the soft set theory , like , soft dens sets , soft codense set , soft nowhere dense set , soft isolated point , soft Baire space ...

Definition 3 .1.1

Let $(\tilde{\chi}_A, \tilde{\tau}, A)$ be a soft topological space ,a soft set F_A is said to be soft dense set iff $Cl(F_A) = \tilde{\chi}_A$.

Example 3 .1.2

If we consider the following universal set χ :

$\chi = \{x_1, x_2. x_3, x_4\}$ and the set of parameters $A \subseteq E$ is $A = \{a_1, a_2, a_3\}$,such that :

$\tilde{\tau} = \{\widetilde{\Phi}_A , \tilde{\chi}_A, F_{1_A}, F_{2_A}\}$, be a soft topological space where

$F_{1_A} = \{(a_1, \{x_1, x_2. x_3\}), (a_2, \{x_1, x_4\}), (a_3, \varphi)\}$

$F_{2_A} = \{(a_1, \{x_1, x_2\}), (a_2, \{x_1, x_4\}), (a_3, \varphi)\}$

Note that $C(\tilde{\chi}) = \{ \widetilde{\Phi}_A , \tilde{\chi}_A , \tilde{\chi}_A - F_{1_A}, \tilde{\chi}_A - F_{2_A} \}$ where :

$\tilde{\chi}_A - F_{1_A} = \{(a_1, \{x_4\}, (a_2, \{x_2, x_3\}), (a_3, \chi)\}$

$\tilde{\chi}_A - F_{2_A} = \{(a_1, \{x_3, x_4\}), (a_2, \{x_2, x_3\}), (a_3, \chi)\}$.

If we consider the soft set $F_A = \{\{(a_1, \{x_1, x_2. x_3\}), (a_2, \{x_2. x_3, x_4\}), (a_3, \varphi)\}$, then , since the smallest soft closed set containing F_A is $\tilde{\chi}_A$, hence by [definition 2.1.20] $Cl(F_A) = \tilde{\chi}_A$ and by the above definition F_A is soft dense set .

Theorem 3.1.3

Let $(\tilde{\chi}_A, \tilde{\tau}, A)$ be a soft topological space ,and let F_A be a soft set over the universe χ, for any soft point a_F of F_A , the following statements are equivalent .

(1) F_A is soft dense set .
(2) $extF_A = \tilde{\Phi}_A$
(3) $\tilde{\chi}_A = [ext(\tilde{\chi}_A - F_A) \, \tilde{\cup} \, Sb(\tilde{\chi}_A - F_A)]$
(4) $Cl(\tilde{\chi}_A - F_A) \, \tilde{\cap} \, (\tilde{\chi}_A - Sb(F_A)) = \tilde{\Phi}_A$
(5) $ext(F_A) \, \tilde{\subseteq} \, IntF_A$
(6) $Cl(\tilde{\chi}_A - F_A) \, \tilde{\subseteq} \, Sb(F_A)$

Morover each of these implies to the condition :

(7) $\tilde{\chi}_A = F_A \, \tilde{\cup} \, Sb(F_A)$

Proof (1)// $1 \Longleftrightarrow 2$

Since $\tilde{\chi}_A - ClF_A = \tilde{\chi}_A - \tilde{\chi}_A$ iff $\tilde{\Phi}_A = \tilde{\chi}_A - ClF_A = Int(\tilde{\chi}_A - F_A)$ iff $ext(F_A) = \tilde{\Phi}_A$.

$2 \Rightarrow 3$

Since $ext(F_A) = \tilde{\Phi}_A$, then $\tilde{\chi}_A = IntF_A \, \tilde{\cup} \, Sb(F_A)$ [Theorem 2.2.21part (2)]

$$= ext(\tilde{\chi}_A - F_A) \, \tilde{\cup} \, (\tilde{\chi}_A - Sb(F_A)) = \tilde{\chi}_A$$

$3 \Rightarrow 1$

since $ClF_A = IntF_A \, \tilde{\cup} \, Sb(F_A)$

$$= [\, ext(\tilde{\chi}_A - F_A) \, \tilde{\cup} \, Sb(F_A)] = ext(\tilde{\chi}_A - F_A) \, \tilde{\cup} \, Sb(\tilde{\chi}_A - F_A)$$

$3 \Rightarrow 4$

Is direct , since $[ext(\tilde{\chi}_A - F_A) \, \tilde{\cup} \, Sb(F_A)]^{\,C}$

$$= [\tilde{\chi}_A - [ext(\tilde{\chi}_A - F_A) \, \tilde{\cup} \, [\tilde{\chi}_A - Sb(F_A)] = \tilde{\Phi}_A$$

$$= (\tilde{\chi}_A - IntF_A) \, \tilde{\cap} \, (\tilde{\chi}_A - Sb(F_A))$$

$$= Cl(\tilde{\chi}_A - F_A) \,\tilde{\cap}\, (\tilde{\chi}_A - Sb(F_A)) = \widetilde{\Phi}_A$$

$4 \Rightarrow 5$

Since $Cl(\tilde{\chi}_A - F_A) \,\tilde{\cap}\, (\tilde{\chi}_A - Sb(F_A)) = \widetilde{\Phi}_A$ iff

$(\tilde{\chi}_A - IntF_A) \,\tilde{\cap}\, (\tilde{\chi}_A - Sb(F_A)) = \widetilde{\Phi}_A$ iff $\tilde{\chi}_A - Sb(F_A) \,\tilde{\subseteq}\, IntF_A$ iff

$Int(F_A) \,\tilde{\cup}\, ext(F_A) \,\tilde{\subseteq}\, IntF_A$,then $ext(F_A) \,\tilde{\subseteq}\, IntF_A$.

$5 \Rightarrow 6$

By part (5) $ext(F_A) \,\tilde{\subseteq}\, IntF_A$, then , $\tilde{\chi}_A = IntF_A \,\tilde{\cup}\, Sb(F_A)$ iff

$(\tilde{\chi}_A - IntF_A) \,\tilde{\cap}\, (\tilde{\chi}_A - Sb(F_A)) = \widetilde{\Phi}_A$ iff $Cl(\tilde{\chi}_A - F_A) \,\tilde{\subseteq}\, Sb(F_A)$.

$6 \Rightarrow 1$

Suppose that $Cl(\tilde{\chi}_A - F_A) \,\tilde{\subseteq}\, Sb(F_A)$, that is :

$(\tilde{\chi}_A - IntF_A) \,\tilde{\cap}\, (\tilde{\chi}_A - Sb(F_A)) = \widetilde{\Phi}_A$ iff

$\tilde{\chi}_A = IntF_A \,\tilde{\cup}\, Sb(F_A) \ldots\ldots\ldots\ldots\ldots\ldots. (3.1)$

But $ClF_A = IntF_A \,\tilde{\cup}\, Sb(F_A) \ldots\ldots\ldots\ldots (3.2)$

Hence by (3.1) and (3.2) we have part -1- .

$1 \Leftrightarrow 7$ direct from theorem [2.2.21] ∎ .

Definition 3 .1.4

Let $(\tilde{\chi}_A, \tilde{\tau}, A)$ be a soft topological space ,a soft set F_A is said to be soft codense set iff $\tilde{\chi}_A - F_A$ is soft dense set .

Example 3 .1.5

If we consider [Example *3 .1.2*] , we have $\widetilde{\Phi}_A$ is a soft set over χ , and since $\tilde{\chi}_A - \widetilde{\Phi}_A$ is soft dense set ,[$Cl(\tilde{\chi}_A) = \tilde{\chi}_A$ [definition. 3 .1.3]

Thus $\widetilde{\Phi}_A$ is soft codense set

Definition 3 .1.6

A soft set F_A is said to be crowded (soft dense in itself) iff it has no soft isolated points .

Definition 3 .1.7

Let $(\tilde{\chi}_A, \tilde{\tau}, A)$ be a soft topological space ,a soft set F_A is said to be soft nowhere dense iff $Int\big(Cl(F_A)\big) = \widetilde{\Phi}_A$.

Theorem 3 .1.8

Let $(\tilde{\chi}_A, \tilde{\tau}, A)$ be a soft topological space , a soft set F_A is soft dense in $\tilde{\chi}_A$ iff for each non-null soft open set G_A , then $F_A \tilde{\cap} G_A \neq \widetilde{\Phi}_A$.

Proof //

Suppose that F_A is soft dense set and $\widetilde{\Phi}_A \neq G_A \widetilde{\in} \tilde{\tau}$ such that if possible $F_A \tilde{\cap} G_A = \widetilde{\Phi}_A$, let $a_F \widetilde{\in} G_A \widetilde{\subseteq} \tilde{\chi}_A$ imply that $a_F \widetilde{\in} \tilde{\chi}_A$, and since F_A is soft dense in $\tilde{\chi}_A$,that is $Cl(F_A) = \tilde{\chi}_A$, so $a_F \widetilde{\in} Cl(F_A)$, but $F_A \tilde{\cap} G_A = \widetilde{\Phi}_A$ is a contradiction . hence $F_A \tilde{\cap} G_A \neq \widetilde{\Phi}_A$.

Conversely :

Suppose that the condition holds : let $a_F \widetilde{\in} \tilde{\chi}_A$, and G_A be a soft open set containing a_F , by the condition $F_A \tilde{\cap} G_A \neq \widetilde{\Phi}_A$, thus by [Theorem 2.1.21] $a_F \widetilde{\in} Cl(F_A)$ that is $\tilde{\chi}_A \widetilde{\subseteq} Cl(F_A)$ and hence F_A is soft dense in $\tilde{\chi}_A$. ∎

Theorem 3 .1.9

Let $(\tilde{\chi}_A, \tilde{\tau}, A)$ be a soft topological space ,and let F_A be a soft set over the universe χ ,then the following statements are equivalent :

(1) F_A is soft nowhere dense .
(2) $Cl(F_A)$ contains no soft open sets .
(3) $\tilde{\chi}_A - Cl(F_A)$ is soft dense in $\tilde{\chi}_A$.

Proof//

$1 \Leftrightarrow 2$ since F_A is soft nowhere dense , $Int(Cl(F_A)) = \widetilde{\Phi}_A$.

Hence $Cl(F_A)$ has no soft interior points , this means that for any soft point a_F of $Cl(F_A)$, then there is no soft open set G_A such that :

$a_F \widetilde{\in} G_A \widetilde{\subseteq} Cl(F_A)$, that is $Cl(F_A)$ contains no soft open set .

$2 \Leftrightarrow 3$ we have $Cl(F_A)$ containing no soft open set iff for any soft open set G_A of $a_F \widetilde{\in} \tilde{\chi}_A$, then $G_A \widetilde{\nsubseteq} Cl(F_A)$ thus , $G_A \widetilde{\cap} (\tilde{\chi}_A - Cl(F_A)) \neq \widetilde{\Phi}_A$ iff every soft open set contains a soft point a_F contains a point of $\tilde{\chi}_A - Cl(F_A)$,for any $a_F \widetilde{\in} \tilde{\chi}_A$ iff every $a_F \widetilde{\in} \tilde{\chi}_A$ then $a_F \widetilde{\in} \tilde{\chi}_A - Cl(F_A)$ iff $Cl(\tilde{\chi}_A - Cl(F_A)) = \tilde{\chi}_A$ iff $\tilde{\chi}_A - Cl(F_A)$ is soft dense .

$3 \Leftrightarrow 1$ since $\tilde{\chi}_A - Cl(F_A)$ is soft dense , by [Definition3.1.1] $Cl(\tilde{\chi}_A - Cl(F_A)) = \tilde{\chi}_A$ iff $\tilde{\chi}_A - Int(\tilde{\chi}_A - \tilde{\chi}_A Cl(F_A)) = \widetilde{\Phi}_A$, iff $Int(Cl(F_A)) = \widetilde{\Phi}_A$. Hence F_A is soft nowhere dense . ∎

Note 3 .1.10

Let $(\tilde{\chi}_A, \tilde{\tau}, A)$ be a soft topological space ,and let $a \in A$ and $x \in \chi$,then for a soft set F_A over the universe χ , the following statements are equivalent :

(1) F_A is a −soft nowhere dense .
(2) $a - Cl(F_A)$ contains no $a -$ soft open set .
(3) $\tilde{\chi}_A - (a - Cl(F_A))$ is a −soft dense in $\tilde{\chi}_A$.

Theorem 3 .1.11

The soft union of a finite number soft nowhere dense sets is a soft nowhere dense .

Proof//

It is suffices to prove the theorem for the case of two soft nowhere dense sets .

Let F_A and G_A be two soft nowhere dense sets , then we put :

$H_A = Int(Cl(F_A \widetilde{\cup} G_A))$, so that , $H_A \widetilde{\subseteq} Cl(F_A \widetilde{\cup} G_A) = Cl(F_A) \widetilde{\cup} Cl(G_A)$, it follows that $H_A \widetilde{\cap} (\tilde{\chi}_A - Cl(G_A)) \widetilde{\subseteq} (Cl(F_A) \widetilde{\cup} Cl(G_A)) \widetilde{\cap} (\tilde{\chi}_A - Cl(G_A))$

$$= [Cl(F_A) \widetilde{\cap} (\tilde{\chi}_A - Cl(G_A))] \widetilde{\cup} [Cl(G_A) \widetilde{\cap} (\tilde{\chi}_A - Cl(G_A))]$$

$Cl(F_A) \,\widetilde{\cap}\, (\tilde{\chi}_A - Cl(G_A)) \,\widetilde{\subseteq}\, Cl(F_A)$, thus

$Int\left(H_A \,\widetilde{\cap}\, (\tilde{\chi}_A - Cl(G_A))\right) \,\widetilde{\subseteq}\, Int(Cl(F_A)) = \tilde{\Phi}_A$ [F_A is soft nowhere dense] ,

But $\quad Int\left(H_A \,\widetilde{\cap}\, (\tilde{\chi}_A - Cl(G_A))\right) = H_A \,\widetilde{\cap}\, \tilde{\chi}_A - Cl(G_A)$ [$H_A \,\widetilde{\cap}\, \chi(\tilde{\chi}_A - Cl(G_A))$ is soft open] .

Hence $H_A \,\widetilde{\cap}\, \tilde{\chi}_A - Cl(G_A) = \tilde{\Phi}_A$, that is mean $H_A \,\widetilde{\subseteq}\, Cl(G_A)$, this implies that $H_A \,\widetilde{\subseteq}\, Int(Cl(G_A)) = \tilde{\Phi}_A$ [G_A is soft nowhere dense] , but $H_A = Int(Cl(F_A \,\widetilde{\cup}\, G_A)) = \tilde{\Phi}_A$.

Hence $F_A \,\widetilde{\cup}\, G_A$ is soft nowhere dense . So , by induction we can prove the theorem for any finite number ∎.

Corollary 3 .1.12

The a −soft union of a finite number an a −soft nowhere dense sets is $a - $ soft nowhere dense .

Theorem 3 .1.13

Let $(\tilde{\chi}_A, \tilde{\tau}, A)$ be a soft topological space ,and let F_A be a soft set over the universe χ ,then the following statements are equivalent :

(1) F_A is soft dense set .
(2) The only soft closed set containing F_A is $\tilde{\chi}_A$
(3) The only soft open set disjoint from F_A is $\tilde{\Phi}_A$.
(4) F_A intersect every non-empty soft open set .

Proof// $1 \Longrightarrow 2$ we have F_A is a soft dense , that is $Cl(F_A) = \tilde{\chi}_A$.

Suppose that if possible G_A is a soft closed set , such that $G_A \,\widetilde{\subseteq}\, \tilde{\chi}_A$ and $F_A \,\widetilde{\subseteq}\, G_A$, so $Cl(F_A) \,\widetilde{\subseteq}\, Cl(G_A)$, but G_A is a soft closed set , then [by proposition 2.1.27 part 2] $Cl(G_A) = G_A$, that is $\tilde{\chi}_A \,\widetilde{\subseteq}\, G_A$, [F_A is soft dense] , thus $\tilde{\chi}_A = G_A$,, hence the only soft closed set containing F_A is $\tilde{\chi}_A$.

$2 \Rightarrow 3$ Let G_A be a non-null soft open set such that $F_A \tilde{\cap} G_A = \tilde{\Phi}_A$, so $Cl(F_A) \tilde{\cap} G_A \tilde{\subseteq} Cl(F_A \tilde{\cap} G_A) = \tilde{\Phi}_A$ [by Theorem 2.1.24] ,is impleis that $G_A = \tilde{\Phi}_A$.

$3 \Rightarrow 4$ let G_A be a non-empty soft open set such that $F_A \tilde{\cap} G_A = \tilde{\Phi}_A$, by (2) $G_A = \tilde{\Phi}_A$ [a contradiction with the assumption that $G_A \neq \tilde{\Phi}_A$, thus $F_A \tilde{\cap} G_A \neq \tilde{\Phi}_A$.

$4 \Rightarrow 1$ directly from Theorem [3.1.8] .

Note 3 .1.14

Let $(\tilde{\chi}_A, \tilde{\tau}, A)$ be a soft topological space ,and let $a \in A$ and $x \in \chi$,then for a soft set F_A over the universe χ , the following statements are equivalent :

(1) F_A is a −soft dense set .
(2) The only soft closed set containing F_A at a is $\tilde{\chi}_A$
(3) The only a −soft open set of F_A is $\tilde{\Phi}_a$.
(4) F_A intersect every non-empty a −soft open set .

Theorem 3.1.15

Let $(\tilde{\chi}_A, \tilde{\tau}, A)$ be a soft topological space ,and let F_A be a soft set over the universe χ , then the following statements hold:

1) $Int(F_A)$ is a soft dense iff $\tilde{\chi}_A - F_A$ is soft nowhere dense .
2) $Int(\tilde{\chi}_A - F_A)$ is soft dense iff F_A is soft nowhere dense .
3) F_A is soft nowhere dense iff $\tilde{\chi}_A - Cl(F_A)$ is soft dense .
4) if F_A is soft nowhere dense then $\tilde{\chi}_A - F_A$ is soft dense .

proof//(1)

$Int(F_A)$ is a soft dense iff $Cl(Int(F_A)) = \tilde{\chi}_A$, [Definition 3 .1.1] iff $\tilde{\chi}_A - Cl(Int(F_A)) = \tilde{\Phi}_A$ iff $Int(Cl(\tilde{\chi}_A - F_A)) = \tilde{\Phi}_A$ iff $\tilde{\chi}_A - F_A$ is soft nowhere dense .

proof//(2)

$Int(\tilde{\chi}_A - F_A)$ is soft dense iff $Cl(Int(\tilde{\chi}_A - F_A)) = \tilde{\chi}_A$ iff $\tilde{\chi}_A - Int(Cl(F_A)) = \tilde{\chi}_A$, iff $Int(Cl(F_A)) = \tilde{\Phi}_A$ iff F_A is soft nowhere dense .

proof//(3)

Since F_A is soft nowhere dense , thus $Int(Cl(F_A) = \widetilde{\Phi}_A$,iff $\tilde{\chi}_A - Int(Cl(F_A)) = \tilde{\chi}_A$ iff $Cl(\tilde{\chi}_A - Cl(F_A)) = \tilde{\chi}_A$ iff $\tilde{\chi}_A - Cl(F_A)$ is soft dense

proof//(4)

F_A is soft nowhere dense iff $\tilde{\chi}_A - F_A$ is soft dense , I .e $Int(Cl(F_A) = \widetilde{\Phi}_A$. then $Cl(Int((\tilde{\chi}_A - F_A)) = \tilde{\chi}_A$ and $Int(\tilde{\chi}_A - F_A) \tilde{\subseteq} \tilde{\chi}_A - F_A$, so $\tilde{\chi}_A = Cl(((\tilde{\chi}_A - F_A))$. ∎

Note 3 .1.16

Let $(\tilde{\chi}_A, \tilde{\tau}, A)$ be a soft topological space ,and let $a \in A$ and $x \in \chi$,then for a soft set F_A over the universe χ , the following statements are equivalent :

(1) $a - Int(F_A)$ is an $-$ soft dense iff $\tilde{\chi}_A - F_A$ is a $-$soft nowhere dense .
(2) $a - Int(\tilde{\chi}_A - F_A)$ a $-$soft dense iff F_A is a $-$soft nowhere dense .
(3) F_A is a $-$soft nowhere dense iff $a - Cl(\tilde{\chi}_A - F_A)$ is a $-$soft dense .
(4) if F_A is a $-$soft nowhere dense then $\tilde{\chi}_A - F_A$ is a $-$soft dense .

Note 3 .1.17

Every soft subset of a soft nowhere dense set is soft nowhere dense .

Theorem 3 .1.18

Let $(\tilde{\chi}_A, \tilde{\tau}, A)$ and $(\tilde{\chi}_A, \tilde{\rho}, A)$ are two soft topological spaces , and

$f_{pu}(\tilde{\chi}_A, \tilde{\tau}, A) \rightarrow (\tilde{\chi}_A, \tilde{\rho}, A)$ be a soft homeomorphism map , then a soft set F_A over χ is soft dense in $\tilde{\tau}$ iff $f_{pu}(F_A)$ is soft dense in $\tilde{\rho}$.

Proof//

Let $H_B \tilde{\in} \tilde{\rho}$, then ${f_{pu}}^{-1}(H_B) \tilde{\in} \tilde{\tau}$ [f_{pu} is soft continuous] , so by [Theorem 3 .1.8] $F_A \tilde{\cap} {f_{pu}}^{-1}(H_B) \neq \widetilde{\Phi}_A$ thus $f_{pu}(F_A) \tilde{\cap} H_B \neq \widetilde{\Phi}_A$ therefore $f_{pu}(F_A)$ is soft dense in $\tilde{\rho}$

Conversely :

Suppose that $f_{pu}(F_A)$ is soft dense in $\tilde{\rho}$ and let $G_A \tilde{\in} \tilde{\tau}$, since f_{pu} is soft open map , so $f_{pu}(G_A) \tilde{\in} \tilde{\rho}$ and by [Theorem 3 .1.8] $f_{pu}(F_A) \tilde{\cap} f_{pu}(G_A) \neq \tilde{\Phi}_A$ and $f_{pu}(F_A \tilde{\cap} G_A) \neq \tilde{\Phi}_A$ and so $f_{pu}^{-1}\left(f_{pu}(F_A \tilde{\cap} G_A)\right) \neq \tilde{\Phi}_A$ that is $F_A \tilde{\cap} G_A \neq \tilde{\Phi}_A$.

Therefore F_A is soft dense in $\tilde{\tau}$. ∎

Theorem 3 .1.19

Let $(\tilde{\chi}_A, \tilde{\tau}, A)$ be a soft topological space ,and let $Y \subseteq \chi$, and let $(\tilde{Y}_A, \tilde{\tau}_Y, A)$ be a soft subspace of $(\tilde{\chi}_A, \tilde{\tau}, A)$, if the soft subset F_A of $\tilde{Y}_A$ is soft dense in $\tilde{\tau}$, then F_A is soft dense in $\tilde{\tau}_Y$.

Proof//

Let $F_A \tilde{\in} \tilde{Y}_A$, such that F_A is soft dense in $\tilde{\chi}_A$, then $Cl(F_A) = \tilde{\chi}_A$, so $Cl(F_A) \tilde{\cap} \tilde{Y}_A = \tilde{\chi}_A \tilde{\cap} \tilde{Y}_A = \tilde{Y}_A$ then , $Cl_{\tilde{Y}_A}(F_A) = \tilde{Y}_A$, hence F_A is soft dense in $\tilde{Y}_A$. ∎

Theorem 3 .1.20

Let $(\tilde{\chi}_A, \tilde{\tau}, A)$ be a soft topological space ,and let F_A be a soft set over the universe χ ,if F_A is soft dense set then for any soft open set G_A we have :

$$G_A \tilde{\subseteq} Cl(F_A \tilde{\cap} G_A)$$

Proof//

Assume that G_A be a soft open set .

Case (1) if G_A equal to $\tilde{\Phi}_A$, the result is done .

Case (2) if a_F is a soft point of G_A , I .e $G_A \neq \tilde{\Phi}_A$ and if possible that

$a_F \tilde{\notin} Cl(F_A \tilde{\cap} G_A)$, then by [Theorem 3 .1.8] there is a soft open set H_A of a_F such that $(F_A \tilde{\cap} G_A) \tilde{\cap} H_A = \tilde{\Phi}_A$, put $K_A = G_A \tilde{\cap} H_A$ is a soft open set of a_F[Definition 2.1.1] , hence $F_A \tilde{\cap} K_A = \tilde{\Phi}_A$, but F_A is a soft dense set , a contradiction with [Theorem 3 .1.8] , hence $a_F \tilde{\in} Cl(F_A \tilde{\cap} G_A)$,

and   thus
$G_A \widetilde{\subseteq} Cl(F_A \widetilde{\cap} G_A)$. ∎

Proposition 3 .1.21

Let $(\tilde{\chi}_A, \tilde{\tau}, A)$ be a soft topological space ,and let F_A be a soft set over the universe χ , then the following statements are equivalent :

(1) F_A is soft dense .
(2) for any soft open set G_A , $F_A \widetilde{\cap} G_A \neq \tilde{\Phi}_A$
(3) $\tilde{\chi}_A - Cl(F_A) = \tilde{\Phi}_A$
(4) $Int(\tilde{\chi}_A - F_A) = \tilde{\Phi}_A$
(5) $\tilde{\chi}_A - F_A$ has no non-empty soft open set .

Proof //

$1 \Leftrightarrow 2$ directly by theorem [3 .1.8]

$1 \Rightarrow 3$ since F_A is soft dense set , then $\tilde{\chi}_A = Cl(F_A)$ so $\tilde{\chi}_A - Cl(F_A) = \tilde{\Phi}_A$.

$3 \Leftrightarrow 4$ since $\tilde{\chi}_A - Cl(F_A) = \tilde{\Phi}_A$ iff $Int(\tilde{\chi}_A - F_A) = \tilde{\Phi}_A$ [by proposition 2.1.25part (2)]

$4 \Rightarrow 5$ if possible that $\tilde{\chi}_A - F_A$, has non-empty soft open set G_A , that is

$G_A \widetilde{\subseteq} \tilde{\chi}_A - F_A$, implies that $G_A \widetilde{\subseteq} Int(\tilde{\chi}_A - F_A)$, hence $Int(\tilde{\chi}_A - F_A) \neq \tilde{\Phi}_A$, which is a contradiction with (4) , therefore $\tilde{\chi}_A - F_A$ has no non-empty soft open set .

$5 \Rightarrow 4$ since $Int(\tilde{\chi}_A - F_A) = \widetilde{\cup} \{G_A \widetilde{\in} \tilde{\tau} , G_A \widetilde{\subseteq} \tilde{\chi}_A - F_A\}$, but by (5) $\tilde{\chi}_A - F_A$ has no non-empty soft open set , that is $\tilde{\Phi}_A$ is the only soft open set contained in $\tilde{\chi}_A - F_A$ hence $Int(\tilde{\chi}_A - F_A) = \tilde{\Phi}_A$.

$4 \Rightarrow 1$ since $Int(\tilde{\chi}_A - F_A) = \tilde{\chi}_A - Cl(F_A) = \tilde{\Phi}_A$ by [corollary 2.1.25 part 2] , $\tilde{\chi}_A - (\tilde{\chi}_A - Cl(F_A)) = \tilde{\chi}_A$, that is $Cl(F_A) = \tilde{\chi}_A$. ∎

This theorem is true if we considered the point $a \in A$.

Theorem 3 .1.22

Let $(\tilde{\chi}_A, \tilde{\tau}, A)$ be a soft topological space , for soft set H_A the following statements are equivalent :

(1) H_A is soft dense
(2) If G_A is any soft closed set , and $H_A \tilde{\subseteq} G_A$ then , $G_A = \tilde{\chi}_A$.
(3) Each non-null proper soft open set K_A such that $K_A \tilde{\cap} H_A \neq \tilde{\Phi}_A$.
(4) $Int(\tilde{\chi}_A - G_A) = \tilde{\Phi}_A$
Proof //

$1 \Rightarrow 2$ let H_A be a soft dense set, and let G_A be a soft closed set such that $H_A \tilde{\subseteq} G_A$, then by [proposition.2.1.22 part(2)] $Cl(H_A) \tilde{\subseteq} Cl(G_A) = G_A$, and hence $G_A = \tilde{\chi}_A$.

$2 \Rightarrow 3$ let K_A be any non-null proper soft open set , such that $K_A \tilde{\cap} H_A = \tilde{\Phi}_A$, which implies that $H_A \tilde{\subseteq} \tilde{\chi}_A - H_A$, but $\tilde{\chi}_A - H_A$ is soft closed set . which is a contradiction with (2) , hence $(K_A \tilde{\cap} H_A \neq \tilde{\Phi}_A)$.

$3 \Rightarrow 4$ assume that $Int(\tilde{\chi}_A - G_A) \neq \tilde{\Phi}_A$, now since $Int(\tilde{\chi}_A - G_A)$ is a soft open set , then by [Theorem 2.1.5] $Int(\tilde{\chi}_A - G_A) is\ a\ soft\ open$ nhd at each of its points , then there is a non-null proper soft open set K_A such that : $K_A \tilde{\subseteq} \tilde{\chi}_A - G_A$, and since $Int(\tilde{\chi}_A - G_A) \tilde{\subseteq} \tilde{\chi}_A - G_A$ [proposition 2.1.17 part (2)] , hence $K_A \tilde{\cap} G_A = \tilde{\Phi}_A$, which a contradiction with the assumption , therefore $Int(\tilde{\chi}_A - G_A) = \tilde{\Phi}_A$.

$4 \Rightarrow 1$ since $Int(\tilde{\chi}_A - G_A) = \tilde{\Phi}_A$, so we have that $\tilde{\chi}_A - Int(\tilde{\chi}_A - G_A) = \tilde{\chi}_A$, therefore , we get that , $Cl(G_A) = \tilde{\chi}_A$ and hence G_A is a soft dense . ∎

Definition 3 .1.23

A soft topological space $(\tilde{\chi}_A, \tilde{\tau}, A)$ is said to be soft Baire space , if the soft intersection of each countable family of a soft open soft dense sets is soft dense .

Theorem 3 .1.24

Let $(\tilde{\chi}_A, \tilde{\tau}, A)$ be a soft Baire space , if $\{F_{n_A}, n \in Z^+\}$ is any countable soft closed family of a soft covering of $\tilde{\chi}_A$, then at least one of F_{n_A} must contains a soft open set , that is :

$Int(F_{n_A}) \neq \widetilde{\Phi}_A$ for at least one n .

Proof//

From $\tilde{\chi}_A = \bigcup_1^\infty F_{n_A}$ [the family form a cover of $\tilde{\chi}_A$, that is $\tilde{\chi}_A \tilde{\subseteq} \bigcup_1^\infty F_{n_A}$] . it follows that $\tilde{\chi}_A - \bigcup_1^\infty F_{n_A} = \bigcap_1^\infty(\tilde{\chi}_A - F_{n_A}) = \widetilde{\Phi}_A$, since $(\tilde{\chi}_A, \tilde{\tau}, A)$ is a soft Baire space , not every $(\tilde{\chi}_A - F_{n_A})$ can be a soft dense , consequently , $Cl(\tilde{\chi}_A - F_{n_A}) \neq \tilde{\chi}_A$, for at least one n . that is $Int(F_{n_A}) = \tilde{\chi}_A - (Cl(\tilde{\chi}_A - F_{n_A}) \neq \widetilde{\Phi}_A$. ■

Definition 3 .1.25

Let $(\tilde{\chi}_A, \tilde{\tau}, A)$ be a soft topological space then :

(1) any countable union of a soft nowhere dense sets is called a set of first category or (soft meager) .
(2) any soft set is not of the soft first category is called soft second category or (non-soft category) .
(3) the soft complement of a soft first category is called soft residual

.

Theorem 3 .1.26

In a soft Baire space $(\tilde{\chi}_A, \tilde{\tau}, A)$ a set of the soft first category has no soft interior .

Proof//

Let $\{G_{n_A}, n \in Z^+\}$ be a soft countable family of a soft nowhere dense sets , and let F_A be any soft open set , such that :

$F_A \tilde{\subseteq} \bigcup_1^\infty G_{n_A}$, we will prove that $F_A = \widetilde{\Phi}_A$.

From $F_A \tilde{\subseteq} \bigcup_1^\infty G_{n_A} \tilde{\subseteq} \bigcup_1^\infty \left(Cl(G_{n_A})\right)$, we find :

$\bigcap_1^\infty(\tilde{\chi}_A - Cl(G_{n_A})) \widetilde{\subseteq} \tilde{\chi}_A - F_A$, and by [Theorem 3 .1.15 part (3)] $\tilde{\chi}_A - Cl(G_{n_A})$ be soft dense sets and is soft open , but $\tilde{\chi}_A$ is soft Baire space , that is $\tilde{\chi}_A = Cl(\bigcap_1^\infty (\tilde{\chi}_A - Cl(G_{n_A})) \widetilde{\subseteq} Cl(\tilde{\chi}_A - F_A) = \tilde{\chi}_A - F_A$ imply that $\tilde{\chi}_A = \tilde{\chi}_A - F_A$ and hence $F_A = \tilde{\Phi}_A$. ∎

(3.2) Soft resolvability and soft irresolvability

In this section , we generalize several important concepts in general topology as ; soft resolvable spaces , soft irrsolvable spaces , soft hereditary irrsolvable , soft submaximal , soft hyper connected ...etc , with some theorems and results .

Definition 3.2.1

A soft topological space $(\tilde{\chi}_A, \tilde{\tau}, A)$ is said to be soft resolvable space iff there exist two soft dense sets F_A and G_A such that :

$F_A \widetilde{\cup} G_A = \tilde{\chi}_A$ and $F_A \widetilde{\cap} G_A = \tilde{\Phi}_A$. otherwise $\tilde{\chi}_A$ is called soft irresolvable space . (I . e for each two soft dense sets F_A and G_A , then $F_A \widetilde{\cap} G_A \neq \tilde{\Phi}_A$.

Note 3.2.2

Let $Y \subseteq \chi$, a soft subspace $(\tilde{Y}_A, \tilde{\tau}_Y , A)$ is soft resolvable subspace if there exist two soft dense soft subsets of $\tilde{Y}_A$ F_{1_A}, F_{2_A} such that $Cl_{\tilde{Y}}(F_{1_A}) = \tilde{Y}_A$ and $Cl_{\tilde{Y}}(F_{2_A}) = \tilde{Y}_A$ with $F_{1_A} \widetilde{\cap} F_{2_A} = \tilde{\Phi}_A$ and $F_{1_A} \widetilde{\cup} F_{2_A} = \tilde{Y}_A$, otherwise $\tilde{Y}_A$ is called soft irresolvable subspace of the space $\tilde{\chi}_A$.

Theorem 3.2.3

Every soft open subset of a soft resolvable space is soft resolvable subspace .

Proof//

Let $(\tilde{\chi}_A, \tilde{\tau}, A)$ be a soft resolvable space , then there exist two disjoint soft dense sets F_A and K_A such that $\tilde{\chi}_A = F_A \widetilde{\cup} K_A$, now let $\tilde{Y}_A$ be a soft open subset of $\tilde{\chi}_A$, so by [Theorem 3.1.20] $\tilde{Y}_A \widetilde{\subseteq} Cl(F_A \widetilde{\cap} \tilde{Y}_A)$ so

$\tilde{Y}_A \tilde{\cap} F_A$ is a soft dense subset of $\tilde{Y}_A$, but $(K_A \tilde{\cap} \tilde{Y}_A) \tilde{\cap} (F_A \tilde{\cap} \tilde{Y}_A) = \tilde{\Phi}_A$ and $K_A \tilde{\cap} \tilde{Y}_A) \tilde{\cup} (F_A \tilde{\cap} \tilde{Y}_A) = \tilde{Y}_A \tilde{\cap} (K_A \tilde{\cup} H_A) = \tilde{Y}_A \tilde{\cap} \tilde{\chi}_A = \tilde{Y}_A$ .Therefore $(\tilde{Y}_A, \tilde{\tau}_Y , A)$ is a soft resolvable subspace of $(\tilde{\chi}_A, \tilde{\tau}, A)$.

Definition 3.2.4

A soft topological space $(\tilde{\chi}_A, \tilde{\tau}, A)$ is said to be soft hereditarily irresolvable if every soft subspace is soft irresolvable .

Note we say that $(\tilde{\chi}_A, \tilde{\tau}, A)$ is soft hereditarily irresolvable iff it does not contain any non-null soft resolvable subspace .

Definition 3.2.5

A soft topological space $(\tilde{\chi}_A, \tilde{\tau}, A)$ is said to be strongly soft irresolvable iff each soft open subspace is soft irresolvable subspace .

Definition 3.2.6

A soft topological space $(\tilde{\chi}_A, \tilde{\tau}, A)$ is said to be soft submaximal iff every soft dense set is soft open .

Definition 3.2.7

A soft topological space $(\tilde{\chi}_A, \tilde{\tau}, A)$ is said to be soft hyper connected iff every soft open set is soft dense .

Example 3.2.8

Consider we have [Example 3 .1.2] , then the soft space is soft hyper connected .

Theorem 3.2.9

For any soft topological space $(\tilde{\chi}_A, \tilde{\tau}, A)$ then the following hold :

Soft Submaximality $\longrightarrow$ soft hereditary irresolvable $\longrightarrow$ strongly soft irresolvable $\longrightarrow$ soft irresolvable

Proof //

Suppose that the space is soft submaximal , and let F_A be any non-null soft resolvable subspace , then there are two soft dense sets G_{1_A} and

G_{2_A} in F_A such that $F_A = G_{1_A} \widetilde{U} G_{2_A}$, then since $\tilde{\chi}_A - G_{2_A} = (\tilde{\chi}_A - F_A) \widetilde{U} G_{1_A}$ is a soft dense in $\tilde{\chi}_A$,[because of $\tilde{\chi}_A - F_A \widetilde{\subseteq} Cl(\tilde{\chi}_A - F_A)$ and $F_A = F_A \widetilde{\cap} Cl(G_{1_A}), then\ \tilde{\chi}_A = (\tilde{\chi}_A - F_A) \widetilde{U} F_A \widetilde{\subseteq} Cl(\tilde{\chi}_A - G_{2_A}) = Cl(\tilde{\chi}_A - F_A) \widetilde{U} Cl(G_{1_A})]$ by the soft submaximality of $\tilde{\chi}_A$ then , $\tilde{\chi}_A - G_{2_A} = (\tilde{\chi}_A - F_A) \widetilde{U} G_{1_A}$ is soft open , so $\tilde{\chi}_A - (\tilde{\chi}_A - G_{2_A}) = G_{2_A}$ is soft closed , and $F_A \widetilde{\subseteq} Cl(G_{2_A}) = G_{2_A}$ [because of $F_A = F_A \widetilde{\cap} Cl(G_{2_A})$] , this implies that $G_{1_A} = \widetilde{\Phi}_A$, [because $F_A = G_{1_A} \widetilde{U} G_{2_A}$] and since $Cl(G_{1_A}) = F_A$, then this contradiction shows that the soft space is soft hereditary irresolvable .

Now , if the soft space is soft hereditary irresolvable , by [Definition 3.2.4] all the soft proportional spaces that are open are soft irresolvable , so that $\tilde{\chi}_A$ is strongly soft irresolvable .

If $\tilde{\chi}_A$ is soft strongly irresolvable , then it is certainly soft irresolvable . ∎

Theorem 3.2.10

A soft topological space $(\tilde{\chi}_A, \tilde{\tau}, A)$ is soft resolvable iff there is a non- null soft dense set F_A such that $Int(F_A) = \widetilde{\Phi}_A$.

Proof //

Since $(\tilde{\chi}_A, \tilde{\tau}, A)$ is soft resolvable , there exists two soft sets F_A and G_A such that : $F_A \widetilde{U} G_A = \tilde{\chi}_A$ and $F_A \widetilde{\cap} G_A = \widetilde{\Phi}_A$, thus $(\tilde{\chi}_A - F_A) = G_A$, but $Cl(F_A) = Cl(G_A) = \tilde{\chi}_A$, hence $Cl(\tilde{\chi}_A - F_A) = \tilde{\chi}_A$ and thus $\tilde{\chi}_A - Cl(\tilde{\chi}_A - F_A) = \widetilde{\Phi}_A$,implies that $Int(F_A) = \widetilde{\Phi}_A$. [corollary 2.1.26] .

Conversely :

Suppose that the condition holds :

Since $Int(F_A) = \widetilde{\Phi}_A$ implies that $\tilde{\chi}_A - Int(F_A) = \tilde{\chi}_A$, that is $Cl(\tilde{\chi}_A - F_A) = \tilde{\chi}_A$ [corollary 2.1.25], thus $F_A \widetilde{U} (\tilde{\chi}_A - F_A) = \tilde{\chi}_A$ and $F_A \widetilde{\cap} \tilde{\chi}_A - F_A = \widetilde{\Phi}_A$ [proposition 1.1.57] , so $Cl(F_A) = Cl(\tilde{\chi}_A - F_A) = \tilde{\chi}_A$ hence $(\tilde{\chi}_A, \tilde{\tau}, A)$ is soft resolvable space . ∎

Theorem 3.2.11

A soft strongly irresolvability of a soft topological space $(\tilde{\chi}_A, \tilde{\tau}, A)$ is equivalent to the following :

(1) every soft open is soft irresolvable subspace .

(2) every soft dense subset has a soft dense interior .

(3) every soft codense set is soft nowhere dense .

(4) every soft subset is a soft union of soft open set and soft nowhere dense .

Proof//$1 \Leftrightarrow 2$ if F_A is a soft dense and if possible $G_A = \tilde{\chi}_A - Cl\big(Int(F_A)\big) \neq \tilde{\Phi}_A$.

Since we have $Int(F_A) \tilde{\subseteq} Cl\big(Int(F_A)\big)$ implying that [proposition 1.1.57 part 5] $\tilde{\chi}_A - Cl\big(Int(F_A)\big) \tilde{\subseteq} \tilde{\chi}_A - Int(F_A)$, that is :

$\tilde{\chi}_A - Cl\big(Int(F_A)\big) \tilde{\cap} \tilde{\chi}_A - Int(F_A) = \tilde{\chi}_A - Cl\big(Int(F_A)\big)$, this means that $G_A \tilde{\cap} Cl(\tilde{\chi}_A - F_A) = G_A \ldots\ldots\ldots\ldots\ldots\ldots . (3.3)$

Now if possible that $G_A - F_A = \tilde{\Phi}_A$ imply that $G_A = F_A$, or $G_A \tilde{\subseteq} F_A$ that is $F_A = \tilde{\chi}_A - Cl\big(Int(F_A)\big)$ and so $\tilde{\chi}_A - Cl\big(Int(F_A)\big) = Int(F_A)$, hence $\tilde{\chi}_A - Cl\big(Int(F_A)\big) \tilde{\subseteq} Int(F_A)$, Now , since $\tilde{\chi}_A - Cl\big(Int(F_A)\big) = \tilde{\chi}_A - Cl\big(Int(F_A)\big) \tilde{\cap} \tilde{\chi}_A - Int(F_A) \tilde{\subseteq} Int(F_A) \tilde{\cap} \tilde{\chi}_A - Int(F_A) = \tilde{\Phi}_A$, impleis that $G_A = \tilde{\chi}_A - Cl\big(Int(F_A)\big) = \tilde{\Phi}_A$, [a contradiction] . hence $G_A - F_A \neq \tilde{\Phi}_A$.

Also $G_A \tilde{\cap} F_A \neq \tilde{\Phi}_A$, if it is not , that is : $G_A \tilde{\subseteq} \tilde{\chi}_A - F_A$, so $\tilde{\chi}_A - Cl\big(Int(F_A)\big) \tilde{\subseteq} \tilde{\chi}_A - F_A$, this implies that $\tilde{\chi}_A = Cl(F_A) \tilde{\subseteq} Cl\big(Int(F_A)\big)$, consequently , $Cl\big(Int(F_A)\big) = \tilde{\chi}_A$.

But $G_A = \tilde{\chi}_A - Cl\big(Int(F_A)\big) = Cl\big(Int(F_A)\big) - Cl\big(Int(F_A)\big) = \tilde{\Phi}_A$, which is a contradiction , thus $G_A \tilde{\cap} F_A \neq \tilde{\Phi}_A$.

Now since F_A is soft dense and $G_A \tilde{\cap} F_A$ is soft and G_A is soft open , then $G_A \tilde{\subseteq} Cl(G_A \tilde{\cap} F_A)$, that is $Cl(G_A \tilde{\cap} F_A) \tilde{\cap} G_A = G_A$, thus $G_A \tilde{\cap} F_A$ is soft dense in G_A.

$Int(\ G_A \,\tilde{\cap}\, F_A) \,\tilde{\subseteq}\, G_A \,\tilde{\cap}\, Int(F_A) = \widetilde{\Phi}_A$ $[F_A$ is soft dense $]$. Thus $G_A - F_A$ is also soft dense in G_A and G_A is soft resolvable , this contradiction shows that $\tilde{\chi}_A - Cl\big(Int(F_A)\big) = \widetilde{\Phi}_A$, and hence $Int(F_A)$ is soft dense .

$2 \Rightarrow 3$ let F_A is soft codense set , then $\tilde{\chi}_A - F_A$ is soft dense , and $Int(\tilde{\chi}_A - F_A)$ is soft dense , so $Int\big(Cl(F_A)\big) = \widetilde{\Phi}_A$.

Hence F_A is soft nowhere dense .

$3 \Rightarrow 4$ case (1): $Int(F_A) = \widetilde{\Phi}_A$, then F_A is soft condense and also $Cl\big(Int(F_A)\big) = \widetilde{\Phi}_A$, so by (3) F_A is soft nowhere dense , hence

$F_A = \widetilde{\Phi}_A \,\tilde{\cup}\, F_A$ is the union of the soft nowhere dense and soft open .

Case (2) : if $Int(F_A) \neq \widetilde{\Phi}_A$, that means we have two subcases :

First : $\tilde{\chi}_A - Int(F_A) = \widetilde{\Phi}_A$ implies that $Int(F_A) = \tilde{\chi}_A$ which means that $F_A = \tilde{\chi}_A$ and so we get the result .

Second : $\tilde{\chi}_A - Int(F_A) \neq \widetilde{\Phi}_A$, so $Cl(\tilde{\chi}_A - \big(F_A \,\tilde{\cap}\, (\tilde{\chi}_A - Int(F_A))\big) =$ $Cl[(\tilde{\chi}_A - F_A) \,\tilde{\cup}\, (\tilde{\chi}_A - \tilde{\chi}_A - Int(F_A)] = Cl[(\tilde{\chi}_A - F_A)] \,\tilde{\cup}\, Cl\big(Int(F_A)\big) =$ $\tilde{\chi}_A - Int(F_A) \,\tilde{\cup}\, Cl\big(Int(F_A)\big) = \tilde{\chi}_A$, hence $F_A - Int(F_A)$ is soft co dense , so by (3) that $F_A - Int(F_A)$ is soft nowhere dense , thus $F_A = Int(F_A) \,\tilde{\cup}\, F_A - Int(F_A)$ is a union of a soft open set and soft nowhere dense .

$4 \Rightarrow 1$ let F_A be a non-null soft open set and resolvable , then $F_A = F_{1_A} \,\tilde{\cup}\, F_{2_A}$, each of F_{i_A} $i = 1,2$ are soft dense in F_A , so $F_{1_A} = H_A \,\tilde{\cup}\, K_A$, where H_A is soft open , and K_A is soft nowhere dense .

If $H_A = \widetilde{\Phi}_A$ imply that $F_A \,\tilde{\subseteq}\, Int\left(Cl(F_{1_A})\right) = Int\big(Cl(K_A)\big) = \widetilde{\Phi}_A$.

So if $H_A \neq \widetilde{\Phi}_A$, $Int_{F_A}\big(F_{1_A}\big) = Int\big(F_{1_A}\big) \neq \widetilde{\Phi}_A$, it implies that $F_{2_A} \,\tilde{\cap}\, Int\big(F_{1_A}\big) \neq \widetilde{\Phi}_A$,a contradiction shows that F_A is soft proportional irresolvable. ■

Theorem 3.2.12

Let F_A has a soft dense interior in a soft topological space $(\tilde{\chi}_A, \tilde{\tau}, A)$, if F_{1_A} and F_{2_A} are two soft dense sets in $\tilde{\chi}_A$ such that $F_{1_A} \tilde{\cap} F_{2_A} \tilde{\cap} F_A = \tilde{\Phi}_A$, then there exist two disjoint soft dense sets G_{1_A} and G_{2_A} of F_A .

Proof //

Since F_{1_A} and F_{2_A} are soft dense sets ,then
$$Int(F_A) \tilde{\cap} Cl(F_{1_A}) \tilde{\subseteq} Cl\left(Int(F_A \tilde{\cap} F_{1_A})\right) \quad \text{and}$$
$$Int(F_A) \tilde{\cap} Cl(F_{2_A}) \tilde{\subseteq} Cl(Int(F_A) \tilde{\cap} F_{2_A}), \text{so} \quad \text{by} \quad \text{hypothesis} \quad \text{that}$$
$$\tilde{\chi}_A = Cl(Int(F_A)) \tilde{\subseteq} Cl(Int(F_A) \tilde{\cap} F_{1_A}) \quad \text{and}$$
$$\tilde{\chi}_A = Cl(Int(F_A)) \tilde{\subseteq} Cl(Int(F_A) \tilde{\cap} F_{2_A}) \quad , \quad \text{so}$$
$$\tilde{\chi}_A \tilde{\subseteq} Cl(Int(F_A) \tilde{\cap} F_{1_A}) \tilde{\subseteq} Cl(F_A \tilde{\cap} F_{1_A})$$
$, \tilde{\chi}_A \tilde{\subseteq} Cl(Int(F_A) \tilde{\cap} F_{2_A}) \tilde{\subseteq} Cl(F_A \tilde{\cap} F_{2_A})$, but $F_{1_A} \tilde{\cap} F_{2_A} \tilde{\cap} F_A = \tilde{\Phi}_A$, hence $F_{1_A} \tilde{\cap} F_A$ and $F_{2_A} \tilde{\cap} F_A$ are soft disjoint subsets of F_A and soft dense sets . ∎

Definition 3.2.13

Let $(\tilde{\chi}_A, \tilde{\tau}, A)$ be a soft topological space . A non-null soft set F_A is called soft proportional resolvable in $\tilde{\chi}_A$ if there are two soft dense sets G_A and H_A with :

$$F_A \tilde{\cap} G_A \neq \tilde{\Phi}_A , F_A \tilde{\cap} H_A \neq \tilde{\Phi}_A \text{ , such that } F_A \tilde{\cap} G_A \tilde{\cap} H_A = \tilde{\Phi}_A .$$

Otherwise F_A is said to be soft proportional irresolvable .

Theorem 3.2.14

Let $(\tilde{\chi}_A, \tilde{\tau}, A)$ be a soft topological space , if the intersection of any two soft dense sets F_A and G_A , is soft dense , then the soft space is soft irresolvable .

proof//

if possible that $\tilde{\chi}_A$ is soft resolvable , then there exist two soft dense sets F_A and G_A in $\tilde{\chi}_A$ such that $F_A \tilde{\cup} G_A = \tilde{\chi}_A$ and $F_A \tilde{\cap} G_A = \tilde{\Phi}_A$, but $Cl(F_A \,\tilde{\cap}\, G_A) = Cl(\tilde{\Phi}_A) = \tilde{\Phi}_A$, a contradiction with the hypotheses . Hence the soft space is soft irresolvable . ∎

Theorem 3.2.15

Let $(\tilde{\chi}_A, \tilde{\tau}, A)$ be a soft topological space . If $\tilde{\chi}_A = F_A \,\tilde{\cup}\, G_A$, and $F_A \,\tilde{\cap}\, G_A = \tilde{\Phi}_A$, where G_A is open and soft hereditary irresolvable and F_A is soft closed and resolvable , then :

(1) If H_A is a soft dense in $\tilde{\chi}_A$, then $G_A \,\tilde{\subseteq}\, Cl(Int(H_A))$.

(2) If U_A is a soft open set , then there exists a soft dense set D_A such that $Int(D_A) \,\tilde{\subseteq}\, G_A \,\tilde{\cup}\, U_A$.

Proof//(1)

Suppose that there exists a soft point $a_F \,\tilde{\in}\, G_A - Cl(Int(H_A))$, that is $a_F \,\tilde{\in}\, G_A$ and $a_F \,\tilde{\in}\, Cl(Int(H_A))$, then choose any soft open set containing a_F say K_A with $K_A \,\tilde{\subseteq}\, G_A$ and $K_A \,\tilde{\cap}\, Int(H_A) = \tilde{\Phi}_A$ that is $K_A \,\tilde{\subseteq}\, Cl(\tilde{\chi}_A - H_A)$, by [Theorem 3.1.20] we have $K_A \,\tilde{\subseteq}\, Cl(K_A \,\tilde{\cap}\, H_A)$ so we get that $K_A = Cl(K_A \,\tilde{\cap}\, H_A) \,\tilde{\cap}\, K_A$, that is $K_A \,\tilde{\cap}\, H_A$ is soft dense in K_A . Also since $K_A \,\tilde{\subseteq}\, Cl(\tilde{\chi}_A - H_A)$, so $K_A = K_A \,\tilde{\cap}\, Cl(\tilde{\chi}_A - H_A)$, then we have by[Theorem 3.1.20]$K_A = [K_A \,\tilde{\cap}\, Cl(\tilde{\chi}_A - H_A)] \,\tilde{\subseteq}\, Cl(K_A \,\tilde{\cap}\, Cl(\tilde{\chi}_A - H_A)))$, then $K_A = [K_A \,\tilde{\cap}\, Cl(K_A \,\tilde{\cap}\, (\tilde{\chi}_A - H_A)]$ implies that $K_A \,\tilde{\cap}\, (\tilde{\chi}_A - H_A)$ is soft dense in K_A , but $(K_A \,\tilde{\cap}\, (\tilde{\chi}_A - H_A)) \,\tilde{\cap}\, (K_A \,\tilde{\cap}\, H_A) = \tilde{\Phi}_A$ hence K_A is soft resolvable subspace of $\tilde{\chi}_A$ which is a contradiction . Hence $G_A \,\tilde{\subseteq}\, Cl(Int(H_A))$.

Proof// (2)

Let U_A is any soft open set , since $\tilde{\chi}_A = G_A \,\tilde{\cup}\, F_A$ and $G_A \,\tilde{\cap}\, F_A = \tilde{\Phi}_A$, so by [Theorem 2.1.24] $F_A \,\tilde{\cap}\, Cl(G_A) \,\tilde{\subseteq}\, Cl(G_A \,\tilde{\cap}\, F_A) = \tilde{\Phi}_A$ and $\tilde{\chi}_A = Cl(G_A) \,\tilde{\cup}\, Cl(F_A) = F_A \,\tilde{\cup}\, Cl(G_A)$, which implies that $\tilde{\chi}_A - Cl(G_A) = F_A$.

Put $W_A = G_A \,\tilde{\cup}\, U_A$ is a soft open set , so $Cl(W_A) = Cl(G_A) \,\tilde{\cup}\, Cl(U_A)$, that is $\tilde{\chi}_A - Cl(W_A) = (\tilde{\chi}_A - Cl(G_A)) \,\tilde{\cap}\, (\tilde{\chi}_A - Cl(U_A))$, thus $\tilde{\chi}_A - Cl(W_A) \,\tilde{\subseteq}\, (\tilde{\chi}_A - Cl(G_A)) = F_A \ldots\ldots\ldots(3.3)$.

But $\tilde{\chi}_A - Cl(W_A)$ is a soft open subset of a soft resolvable subspace F_A , then by [Theorem 3.2.3] $\tilde{\chi}_A - Cl(W_A)$ is soft resolvable subspace of F_A then there exists two disjoint soft dense sets E_{1_A}, E_{2_A} of $\tilde{\chi}_A - Cl(W_A)$ such that $\tilde{\chi}_A - Cl(W_A) = E_{1_A} \, \tilde{\cup} \, E_{2_A}$..(3.4) .

And $\tilde{\chi}_A - Cl(W_A) \tilde{\subseteq} Cl(E_{1_A})$, $\tilde{\chi}_A - Cl(W_A) \tilde{\subseteq} Cl(E_{2_A})$ ....(3.5) .

Put $\qquad\qquad\qquad D_A = W_A \, \tilde{\cup} \, E_{1_A} \qquad\qquad$, $\qquad\qquad$ since $\tilde{\chi}_A = Cl(W_A) \, \tilde{\cup} \, \left(\tilde{\chi}_A - Cl(W_A)\right) \tilde{\subseteq} Cl(W_A) \, \tilde{\cup} \, Cl(E_{1_A}) = Cl(D_A)$, thus D_A is soft dense .

Now since $W_A \tilde{\subseteq} D_A$ and W_A is soft open , so $W_A \tilde{\subseteq} Int(D_A)$ ....(3.6) .

To show that $Int(D_A) \tilde{\subseteq} W_A$.

Let $a_F \tilde{\in} Int(D_A)$, then there exists $V_A \tilde{\in} \tilde{\tau}$ such that

$$V_A \tilde{\subseteq} D_A = W_A \, \tilde{\cup} \, E_{1_A} \,..(3.7) .$$

Then $V_A \tilde{\cap} E_{2_A} = \tilde{\Phi}_A$, if not , we can choose $a_{1_F} \tilde{\in} V_A$ and $a_{1_F} \tilde{\in} E_{2_A}$, so $a_{1_F} \tilde{\notin} E_{1_A}$ and $a_{1_F} \tilde{\in} D_A$, so $a_{1_F} \tilde{\in} W_A$, but from (3.4) that $(\tilde{\chi}_A - Cl(W_A)) = E_{1_A} \, \tilde{\cup} \, E_{2_A}$ we get that $a_{1_F} \tilde{\in} \tilde{\chi}_A - Cl(W_A)$ which means that $W_A \tilde{\cap} (\tilde{\chi}_A - Cl(W_A)) \neq \tilde{\Phi}_A$. Which is a contradiction . Hence $V_A \tilde{\cap} E_{2_A} = \tilde{\Phi}_A$, so $V_A \tilde{\cap} Cl(E_{2_A}) = \tilde{\Phi}_A$, if not then there exists a soft point a_{2_F} such that $a_{2_F} \tilde{\in} V_A$ and $a_{2_F} \tilde{\in} Cl(E_{2_A})$, so $a_{2_F} \tilde{\in} D(E_{2_A})$ hence by [Definition 2.2.1] for each soft open set H_A containing a_{2_F} we have $E_{2_A} \tilde{\cap} (H_A - a_{2_F}) \neq \tilde{\Phi}_A$, but V_A containing a_{2_F} and $V_A \tilde{\cap} E_{2_A} = \tilde{\Phi}_A$ which is a contradiction , that is $V_A \tilde{\cap} Cl(E_{2_A}) = \tilde{\Phi}_A \,.............................(3.8).$

Then from (3.8) we get that $V_A \tilde{\cap} E_{1_A} = \tilde{\Phi}_A$, if not then there exists a soft point $a_{3_F} \tilde{\in} V_A$ and $a_{3_F} \tilde{\in} E_{1_A}$, so from (3.4) that $a_{3_F} \tilde{\in} \tilde{\chi}_A - Cl(W_A)$, and by (3.7) we get that $a_{3_F} \tilde{\notin} Cl(E_{2_A})$.

But from (3.5) $[\tilde{\chi}_A - Cl(W_A)] \tilde{\subseteq} Cl(E_{2_A})$, this is a contradiction , hence $V_A \tilde{\cap} E_{1_A} = \tilde{\Phi}_A \,...(3.9) .$

Therefore , from (3.7) and (3.9) , we have $V_A \cong W_A$ so $a_F \in W_A$, thus $Int(D_A) \cong W_A$.(3.10) .

Hence from (3.6) and (3.10) we get that $Int(D_A) = G_A \widetilde{U} U_A$.

Theorem 3.2.16

Let $(\tilde{\chi}_A, \tilde{\tau}, A)$ be a soft topological space ,Then a soft set F_A over the universe χ with $Int(F_A) \neq \widetilde{\Phi}_A$ is soft proportional resolvable iff for any soft dense set G_A , then $Int(G_A \widetilde{\cap} F_A) \neq \widetilde{\Phi}_A$.

Proof//

Suppose that F_A be a soft set over the universe χ with $Int(F_A) \neq \widetilde{\Phi}_A$ is a soft proportional resolvable , if possible that $Int(G_A \widetilde{\cap} F_A) = \widetilde{\Phi}_A$, for some soft dense G_A .

Since $Int(F_A) \neq \widetilde{\Phi}_A$, $F_A \not\subseteq G_A$,and hence $Cl(G_A) = \tilde{\chi}_A - Int(G_A \widetilde{\cap} F_A)$ is soft dense in $\tilde{\chi}_A$ .with $Cl(G_A) \widetilde{\cap} F_A \neq \widetilde{\Phi}_A$, but $G_A \cap Cl(G_A) \widetilde{\cap} F_A = ((G_A \widetilde{\cap} \tilde{\chi}_A - Cl(G_A)) \widetilde{U} ((\tilde{\chi}_A - F_A) \widetilde{\cap} F_A) = \widetilde{\Phi}_A$, and hence F_A is soft proportional resolvable . a contradiction , hence $Int(G_A \widetilde{\cap} F_A) \neq \widetilde{\Phi}_A$.

Conversely :

Let $Int(G_A \widetilde{\cap} F_A) \neq \widetilde{\Phi}_A$,for any soft dense set G_A in $\tilde{\chi}_A$, where $Int(F_A) \neq \widetilde{\Phi}_A$, let G_{1_A} and G_{2_A} are two soft dense sets , then $G_{1_A} \widetilde{\cap} F_A \neq \widetilde{\Phi}_A$ and $G_{2_A} \widetilde{\cap} F_A \neq \widetilde{\Phi}_A$, thus $G_{1_A} \widetilde{\cap} Int(F_A) \neq \widetilde{\Phi}_A$ and $G_{2_A} \widetilde{\cap} Int(F_A) \neq \widetilde{\Phi}_A$, a gain since $Int(G_{1_A} \widetilde{\cap} F_A) \neq \widetilde{\Phi}_A$, we get that $G_{2_A} \widetilde{\cap} Int(G_{1_A} \widetilde{\cap} F_A) \neq \widetilde{\Phi}_A$, then $Int(G_{1_A} \widetilde{\cap} F_A) \cong G_{1_A} \widetilde{\cap} F_A$, that is $G_{2_A} \widetilde{\cap} G_{1_A} \widetilde{\cap} F_A \neq \widetilde{\Phi}_A$, so F_A is a soft irresolvable in $\tilde{\chi}_A$. ∎

From above , we get the following corollary :

Corollary 3.2.17

Let $(\tilde{\chi}_A, \tilde{\tau}, A)$ be a soft topological space and F_A be any soft set over the universe χ with $Int(F_A) \neq \widetilde{\Phi}_A$, then F_A is soft proportional irresolvable in $\tilde{\chi}_A$ iff $Int(F_A)$ is soft proportional irresolvable in $\tilde{\chi}_A$.

In the following theorem we discuss the situation , when $Int(F_A) = \widetilde{\Phi}_A$.

Theorem 3.2.18

Let $(\tilde{\chi}_A, \tilde{\tau}, A)$ be a soft topological space and F_A be any soft set over the universe χ , with $|F_A| > 1$ [$|F_A|$ is the cardinality of F_A] and $Int(F_A) = \widetilde{\Phi}_A$, then F_A is soft proportional resolvable in $\tilde{\chi}_A$.

Proof//

Let $a_F \widetilde{\in} F_A$, since $|F_A| > 1$, then $F_A - a_F \neq \widetilde{\Phi}_A$, put $G_{1_A} = \tilde{\chi}_A - a_F$ and $G_{2_A} = \tilde{\chi}_A - (F_A - a_F)$ now since $Int(F_A) = \widetilde{\Phi}_A$, so $Cl(\tilde{\chi}_A - F_A) = \tilde{\chi}_A - Int(F_A) = \tilde{\chi}_A$, that is $\tilde{\chi}_A - F_A$ is soft dense , $G_{2_A} = \tilde{\chi}_A - (F_A - a_F) = \tilde{\chi}_A - F_A \widetilde{\cup} a_F$, so $Cl(G_{2_A}) = Cl(F_A - a_F) \widetilde{\cup} Cl(a_F) = \tilde{\chi}_A$ .Hence G_{2_A} is soft dense . Now since $Int(a_F) \widetilde{\subseteq} Int(F_A) = \widetilde{\Phi}_A$, then $Cl(G_{1_A}) = Cl(\tilde{\chi}_A - a_F) = \tilde{\chi}_A - Int(a_F)$, thus G_{1_A} is soft dense set .It is clear that

$G_{1_A} \widetilde{\cap} F_A \neq \widetilde{\Phi}_A$ and $G_{2_A} \widetilde{\cap} F_A \neq \widetilde{\Phi}_A$, but $G_{1_A} \widetilde{\cap} G_{2_A} \widetilde{\cap} F_A = \widetilde{\Phi}_A$, therefore F_A is soft proportional resolvable . ∎

Theorem 3.2.19

Each singleton soft point is soft proportional irresolvable .

Proof//

Let $F_A = \{a_F\}$ is a single soft point and let G_{1_A} , G_{2_A} be two soft dense sets such that $G_{1_A} \widetilde{\cap} F_A \neq \widetilde{\Phi}_A$ and $G_{2_A} \widetilde{\cap} F_A \neq \widetilde{\Phi}_A$, hence $G_{1_A} \widetilde{\cap} F_A = \{a_F\} = G_{2_A} \widetilde{\cap} F_A$, so $G_{1_A} \widetilde{\cap} G_{2_A} \widetilde{\cap} F_A \neq \widetilde{\Phi}_A$, therefore F_A is soft proportional irresolvable . ∎

Theorem 3.3.20

Let $(\tilde{\chi}_A, \tilde{\tau}, A)$ be a soft topological space , If there exists a soft set F_A is a soft proportional irresolvable in $\tilde{\chi}_A$, then $(\tilde{\chi}_A, \tilde{\tau}, A)$ is soft irresolvable .

Proof //

Suppose that F_{1_A}, F_{2_A} are two soft dense sets in $\tilde{\chi}_A$, we must prove that $F_{1_A} \tilde{\cap} F_{2_A} \neq \widetilde{\Phi}_A$, since F_A is a soft proportional irresolvable in $\tilde{\chi}_A$, then $F_A \tilde{\cap} F_{1_A} \neq \widetilde{\Phi}_A$ and $F_A \tilde{\cap} F_{2_A} \neq \widetilde{\Phi}_A$ and $F_{1_A} \tilde{\cap} F_{2_A} \tilde{\cap} F_A \neq \widetilde{\Phi}_A$, but $F_{1_A} \tilde{\cap} F_{2_A} \tilde{\cap} F_A \tilde{\subseteq} F_{1_A} \tilde{\cap} F_{2_A}$ Hence $F_{1_A} \tilde{\cap} F_{2_A} \neq \widetilde{\Phi}_A$ .Thus $(\tilde{\chi}_A, \tilde{\tau}, A)$ is soft irresolvable . ∎

Now we get the following corollary :

Corollary 3.4.21

A soft topological space $(\tilde{\chi}_A, \tilde{\tau}, A)$ is soft irresolvable iff there exists a soft set F_A with $Int(F_A) \neq \widetilde{\Phi}_A$ such that F_A is soft proportional irresolvable in $\tilde{\chi}_A$.

Theorem 3.2.22

Let $(\tilde{\chi}_A, \tilde{\tau}, A)$ be a soft resolvable space , then for any soft set F_A, if $Int(F_A) \neq \widetilde{\Phi}_A$, then F_A is a soft proportional resolvable in $\tilde{\chi}_A$.

Proof //

Since the soft space is soft resolvable , then there are two soft dense sets G_{1_A} and G_{2_A} in $\tilde{\chi}_A$ such that $\tilde{\chi}_A = G_{1_A} \tilde{\cup} G_{2_A}$ and $G_{1_A} \tilde{\cap} G_{2_A} = \widetilde{\Phi}_A$, and let $Int(F_A) \neq \widetilde{\Phi}_A$ and that F_A is a soft open set , so by [Theorem . 3 .1.8] $G_{1_A} \tilde{\cap} Int(F_A) \neq \widetilde{\Phi}_A$ and $G_{2_A} \tilde{\cap} Int(F_A) \neq \widetilde{\Phi}_A$, but $Int(F_A) \tilde{\subseteq} F_A$ hence $G_{1_A} \tilde{\cap} F_A \neq \widetilde{\Phi}_A$ and $G_{2_A} \tilde{\cap} F_A \neq \widetilde{\Phi}_A$ and clearly that $G_{1_A} \tilde{\cap} G_{2_A} \tilde{\cap} F_A = \widetilde{\Phi}_A$ and hence F_A is soft proportional resolvable in $\tilde{\chi}_A$. ∎

References

[1] D. Molodtsov , Soft set theory –first results , computers and mathematics with applications , 37(1999)19-31.

[2] P.K.Maji . R.Biswas and A.R.Roy , Soft set theory , computers and mathematics with applications , 45(2003)555-562.

[3] N.Cagman , F.Cltak and S.Enginolu , Fp-soft set theory and its applications , annals of fuzzy mathematics and informatics , 2(2011)216-226 .

[4] N.Cagman , Serkan Karatas ,Serdar Enginogllu , Soft topology , computers and mathematics with applications , 62(2011)351-358 .

[5] D.Rose ,K . Sizemore and Benthurston , Strongly irresolvable spaces , international journal of mathematics and mathematical sciences , (2006)1-12 .

[6] H.I.Mustafa and F.M.sleim , Soft generalized closed sets with respect to an ideal in soft topological spaces , App. Math . inf . sci . 2(2014)665-671 .

[7] Hewitt , E: Problem of the theoretical topology , Duck Math. 10(1943)309-333 .

[8] Ganster , m.reilly , I.L. Vamanamurthy and M.K , Dense sets and irresolvable spaces ,Researches Math . xxx vi (1987).

[9] F. Freng , etal , Soft sets combined with fuzzy sets and rough sets , atentative approach , soft compacting , 14 (2010) , 899-911.

[10] S . Saraf , Survey or review on soft set theory and development , the SIJ Transaction on computer science engineering and its applications CCSEA , 3(2013)60 .

[11] M.ifran Ali , F.feng ,X. liu W.Keun min and M. shabir , On some new operations in soft set theory , computers and mathematics with applications , 579(2009)1547-1553 .

[12] P.Majumdar and S.K . Samanta , Generalized fuzzy soft sets , computers and mathematics with applications , 59(2010)1425-1432 .

[13] Sh.Modak , Relativization in in resolvability and irresolvability , international mathematical forum , 6(2011)1059-1062 .

[14] A.V Arhangl'sakii , From classic topological invariant to relative topological properties , Sciatica mathematics Jamponicae , 55(2002)153-201 .

[15] L.Genglie , I.L.Reilly and M.K.Vamanamurthy , Dense sets and irresolvable spaces , Ricerche de Mathematical , xxx 1(1987)163-170 .

[16] L.Genglei, W.Huidong and W.Yunxia , Some properties of relative regularity and compactness , journal of mathematics researches ,vol .1 ,1 (2009).

[17] M.S.Sarsakand and H.Z.Hdeib, Relatively strongly normal and relatively normal subspace , international mathematical forum . 5(2010)579-586 .

[18] S.M.Madhumugal pal , Soft matrices , department of applied mathematics with oceanobgy computer programing , journal of uncertain system , 7(2013)254-264 .

[19] P.K.Maji et al , Fuzzy soft set , the journal of fuzzy mathematics 3(2001)2726-2735 .`

[20] H.Aktas and N.Cageman , Soft sets and soft groups , information science , 177(2007)2726-2735 .

[21] Z.Kong et al , The normal parameter reduction of soft sets and its algorithm , computers and mathematics with applications , (2008)13029-13037 .

[22] Z.Kong et al , comment on "A fuzzy soft set theoretic approach to decision making problems , journal of computational and applied mathematics , 223(2009)540-542 .

[23] N.Cageman and S.Enginolgu , Soft set theory and uni-int decision making European , journal of operational research . 207(2010)848-855 .

[24] N.Cagman , F.Cltak and S.Enginolu , Fp-soft set theory and its applications , annals of fuzzy mathematics and informatics , 2(2011)216-226 .

[25] W.K.Min , Note on soft topological spaces computers and mathematics with applications , 62(2011)3524-3028 .

[26] S.Yuksel , Soft regular generalized closed sets in soft topological spaces , Int . journal of math . analysis 8(2014)355-367 .

[27] B. Ahmed and S. Hussain , On some structures of soft topology , mathematical science (2012) .

[28] T.Y.Ozturk and S.Bayrmov , Soft mapping space , the scientific world journal vol. (2014) Article ID 207992 , 8 pages .

[29] I.Zorlutuna and H.Cakus , On continuity on soft mappings , App . Math . Inf . sci ., 9(2015) 403-409 .

[30] A .Aygunoglu and H.Aygun , Some notes on soft topological spaces , journal computer and applied , 21(2012)113-119 .

[31] I.Zorlutuna , M.Akdag , W.K.Min and S.Asmaka , Remarks on soft topological spaces , annals of fuzzy mathematics and informations , 3(2012)171-185 .

 [32] G.Xuechong , Study on central soft sets :definitions and basic operations , preprint submitted to Elsevier (2015) .

[33] E.pPeyghan, B.Samadi and A.Tyebi , about soft topological spaces , preprint .

[34] S.Bayrmov , C.Gundaz and A.Erdem , Soft path connectedness on soft topological spaces , International conference on computational and mathematical methods in science and engineering [MMSE] (2013) 24–27 .

[35] P.K.Maji and A.R.Roy , An application of soft sets in decision making problem , computers and mathematics with applications , 44(2002)1077-1083 .

[36] F.Feng , Y.B.Jun , X.Z.Zhao , Soft semi rings , computers and mathematics with applications , 56(2008)2621-2628 .

[37] D.N.Georgiou and A.C.Megaritis , Soft set theory and topology , PP.Gen. topology. , 1(2014)93-109 .

[38] M.Shabir and M.Naz , On soft topological spaces , computers and mathematics with applications , 61(2011)1780-1799 .

[39] P.Majumdar and S.K.Samanta , On soft mappings ,computers and mathematics with applications , 60(2010)2666-2672 .

[40] S.Yugsel , Soft regular generalizes closed sets in soft topological spaces , Int . journal of Math. Analysis , 8(2014)355-367 .

[41] F.Karaaslan , N.Cagman and S.Engiloglu , Soft lattices , journal of new results in science , 1(2012)5-17 .

[42] A.Kahral and B.Ahmad , Mappings on soft classes , new math. Nat. Comput. 7 no.3 (2011) , 471-481 .

[43] D. Wardowski , On soft mapping and its fixed points , fixed point theory and its applications , (2013)182

[44] S. Hussain and B. Ahmed , Some properties on soft topological spaces , computers and mathematics with applications , (2011)4058-4067 .

[45] K.V.Babitha and J.J.Sunil , Soft sets relations and functions , computers and mathematics with applications , 60(2010)1840-1849 .

Printed by Books on Demand GmbH, Norderstedt / Germany